TOWARDS A SUSTAINABLE HUMAN RIGHT TO WATER

This publication was made possible through the financial support provided by the Silvia W. de Groot Fund in Leiden and Stichting Schilthuisfonds in The Hague.

Human Rights Research Series, Volume 85.

The titles published in this series are listed at the end of this volume.

TOWARDS A SUSTAINABLE HUMAN RIGHT TO WATER

Supporting vulnerable people and protecting water resources

Daphina MISIEDJAN

intersentia
Cambridge – Antwerp – Chicago

Intersentia Ltd
Sheraton House | Castle Park
Cambridge | CB3 0AX | United Kingdom
Tel.: +44 1223 370 170 | Fax: +44 1223 370 169
Email: mail@intersentia.co.uk
www.intersentia.com | www.intersentia.co.uk

Distribution for the UK and Ireland:
NBN International
Airport Business Centre, 10 Thornbury Road
Plymouth, PL6 7 PP
United Kingdom
Tel.: +44 1752 202 301 | Fax: +44 1752 202 331
Email: orders@nbninternational.com

Distribution for Europe and all other countries:
Intersentia Publishing nv
Groenstraat 31
2640 Mortsel
Belgium
Tel.: +32 3 680 15 50 | Fax: +32 3 658 71 21
Email: mail@intersentia.be

Distribution for the USA and Canada:
Independent Publishers Group
Order Department
814 North Franklin Street
Chicago, IL60610
USA
Tel.: +1 800 888 4741 (toll free) | Fax: +1312 337 5985
Email: orders@ipgbook.com

Towards a Sustainable Human Right to Water. Supporting vulnerable people and protecting water resources

ISBN 978-1-78068-612-7 (paperback)
ISBN 978-1-78068-616-5 (PDF)
D/2019/7849/23
NUR 828

British Library Cataloguing in Publication Data. A catalogue record for this book is available from the British Library.

To my mother, Wilma Morsen

ACKNOWLEDGMENTS

Education is without a doubt highly important and I am grateful to have been able to accomplish this milestone. An opportunity like this, many wish for and I wish it to be achievable for them. Especially, for those who do not always see themselves represented in academic spaces.

This Ph.D. journey required me to remain outside of my comfort zone for long periods of time. It challenged me to grow in ways I could not have imagined and brought me to wonderful places. Without the unwavering support of many this process could not have been successfully completed. I would like to express my great appreciation to them by saying thank you.[1]

I would like to express my sincere gratitude to my wonderful thesis advisors. Marleen van Rijswick always had an open door to discuss issues I had with my research or to just have a talk about the regular things in life. Her patient guidance and enthusiastic encouragement were very much appreciated. Antoine Buyse has been there since the beginning of my academic development and it has been wonderful to have his support and useful critique. Their expertise but also their empathy haven been inspiring and I could not have imagined having better supervisors and mentors for my Ph.D. research.

I would like to acknowledge the reading committee members: prof. mr. Karin Arts, prof. mr. Jenny Goldschmidt, prof. dr. Joyeeta Gupta, prof. dr. Louis Kotzé, prof. mr. Remco Nehmelman and prof. dr. Barbara Oomen, for their insightful comments and questions which have allowed me to be critical of my work and help further develop it for future publication.

I would like to thank Utrecht University for all the facilities they provided. The Utrecht Centre for Water, Oceans and Sustainability Law and the Netherlands Institute for Human Rights have provided me an academic home for the past years which I am grateful for. They allowed me to develop my ideas, interact with students and meet great (international) teachers. Among them em. prof. dr. Flinterman who encouraged me to write my first article and prof. dr. Prechal who mentored me during my bachelor studies.

1 As a legal expert, I would however like to state that this is not an exhaustive list.

I would also like to express my appreciation to my colleagues for their feedback, cooperation, stimulating discussions and humor. And I would like to thank others with whom I have had the pleasure to work with during this and other related projects.

During my research visits to Suriname many governmental institutions, organizations and individuals provided me with valuable information and experiences for the purpose of this research by opening up their doors. With special mention of Waterforum Suriname and the employees of SWM, UNDP Suriname and the Lim A Po Institute for Social Studies. A very special thank you goes out to the Silvia W. de Groot Fund for its financial contribution which financed my research activities in Suriname and also to the Stichting Schilthuisfonds for financially contributing to the publication of this thesis.

I am blessed with a very big supportive family and I am grateful to the families Morsen and Misiedjan for their curiosity about my research and their support. My aunts and uncles who stepped in when needed. My cousins who gave me sound advice and were generous with their time. My nieces and nephews who keep me grounded and were the cutest distractions and reminders that, sometimes, you just have to sit down and play.

I thank my friends for their understanding and support. Especially my sister friends for their Sisterhood. Having them stand beside me and sometimes behind me to give me a nudge was highly appreciated.

And finally, words cannot describe how grateful I am to be able to share this milestone with my mother, Wilma Morsen. This Ph.D. journey is as much hers as it is mine and for this reason this book is dedicated to her.

Thanks for all your encouragement!
Grantangi!

CONTENTS

LIST OF ABBREVIATIONS

ACHPR	African Charter on Human and Peoples' Rights
BOG	Bureau Openbare Gezondheidszorg (Public Health Department)
CEDAW	Convention on the Elimination of All Forms of Discrimination Against Women
CESCR	Committee on Economic, Social and Cultural Rights
CRC	Convention on the Rights of the Child
FOB	Fund for Development of the Interior (Stichting Fonds Ontwikkeling Binnenland)
GA	United Nations General Assembly
HRC	Human Rights Council
HRCom	Human Rights Committee
IACHR	Inter-American Court of Human Rights
ICCPR	International Covenant on Civil and Political Rights
ICESCR	International Covenant on Economic, Social and Cultural Rights
ICJ	International Court of Justice
IL	International Law Commission
IUCN	International Union for Conservation of Nature
IWRM	Integrated Water Resources Management
MDGs	Millennium Development Goals
OAS	Organization of American States
PAHO	Pan American Health Organization
PPP	Polluter Pays Principle
SWM	Suriname Watersupply Company (Surinaamse Water Maatschappij)
UDHR	Universal Declaration on Human Rights
UNFCCC	United Nations Framework Convention on Climate Change
UNHCHR	United Nations High Commissioner for Human Rights
UN	United Nations
UNECE	United Nations Economic Commission for Europe
UNEP	United Nations Environment Programme
UNGA	United Nations General Assembly
SDGs	Sustainable Development Goals
WHO	World Health Organization
WCED	World Commission on Environment and Development

LIST OF TABLES

LIST OF FIGURES

TABLE OF INSTRUMENTS

INTERNATIONAL INSTRUMENTS

Charter of the United Nations (adopted 24 October 1945) 1 UNTS XVI.

Stature of the International Court of Justice (adopted 26 June 1945, entered into force 24 October 1945) 33 UNTS 993

Universal Declaration on Human Rights (adopted 10 December 1948) UNGA Res 217 A (III)

International Convention on the Elimination of All Forms of Racial Discrimination (adopted 7 March 1966, entered into force 4 January 1969) 660 UNTS 195.

International Covenant on Economic, Social and Cultural Rights (adopted 16 December 1966, entered into force 3 January 1976) 993 UNTS 3.

International Covenant on Civil and Political Rights (adopted 16 December 1966, entered into force 23 March 1976) 999 UNTS 171.

Optional Protocol to the International Covenant on Civil and Political Rights (adopted 16 December 1966, entered into force 23 March 1976) 999 UNTS 171

Stockholm Declaration on the Human Environment Report of the United Nations Conference on the Human Environment (Stockholm 5–16 June 1972) (16 June 1972) UN DocA/CONF. 48/14.

Convention on the Elimination of All Forms of Discrimination Against Women (adopted 18 December 1979, entered into force 3 September 1981) 1249 UNTS 13.

Convention against Torture and Other Cruel, Inhuman or Degrading Treatment or Punishment (adopted 10 December 1984, entered into force 26 June 1987) 1465 UNTS 85.

Convention on the Rights of the Child (adopted 20 November 1989, entered into force 2 September 1990) 1577 UNTS 3.

Convention on the Law of the Non-Navigational Uses of International Watercourses (adopted on 21 May 1997, entered into force 17 August 2014)

Optional Protocol to the International Covenant on Economic, Social and Cultural Rights (adopted 10 December 2008, entered into force 5 May 2013) UN Doc A/RES/63/118.

REGIONAL INSTRUMENTS

American Declaration on the Rights and Duties of Man (adopted 2 May 1948) OAS Doc OEA/ Ser.L.V/II82. Doc. 6 rev. 1.

African Charter on Human and Peoples' Rights (adopted 27 June 1981, entered into force 21 October 1986) 1520 UNTS 217.

Additional Protocol to the American Convention on Human Rights in the Area of Economic, Social, and Cultural Rights (adopted 17 November 1988, entered into force 16 November 1999) OASTS 69.

NATIONAL INSTRUMENTS, DOCUMENTS AND CASE-LAW

ARGENTINA

- Constitution of the City of Buenos Aires
- Asociación Civil por la Igualdad y la Justicia c/ Gobierno de la Ciudad de Buenos Aires Cámara de Apelaciones en lo Contencioso Administrativo y Tributario (Ciudad Autónoma de Buenos Aires) 18 July 2007115

BANGLADESH

- Constitution of Bangladesh
- Environmental Conservation Act 1995
- Environmental Conservation Rules 1997
- Rabia Bhuiyan v. Ministry of LGRD and others, Appellate Division (Civil Appeal no 118 of 1999), 27 August 2005.

BOTSWANA

- Constitution of the Republic of Botswana
- Water Act of 1968
- Matsipane Mosethanyane & Others v The Attorney-General of Botswana High Court decision MAHLB-000393-09

COLOMBIA

- Political Constitution of Colombia
- General Environmental Law of Colombia

COSTA RICA

- Political Constitution of the Republic of Costa Rica
- General Law on Safe Drinking Water
- Mining Code

SOUTH-AFRICA

- Constitution of the Republic of South Africa
- City of Johannesburg Metropolitan Municipality Water Services By-Laws

- Water Services Act 108 of 1997
- National Water Standards Regulations, Government Gazette no 22355, Notice R509 of 2001
- Regulations Relating to Compulsory National Standards and Measures to Conserve Water
- Mazibuko and Others v City of Johannesburg and Others (Centre on Housing Rights and Evictions, Amicus Curiae) Constitutional Court of South Africa (CCT 39/09) [2009] ZACC 28; 2010 (3) BCLR 239 (CC); 2010 (4) SA 1 (CC) 8 October 2009

INDIA

- Constitution of India
- Vellore Citizens Welfare Forum vs Union Of India & Ors on 28 August, 1996
- Perumatty Grama Panchayat v State of Kerala High Court (Kerala) on 16 December 2003
- Subhash Kumar v State of Bihar and others Supreme Court on 3 October 2012

SURINAME

- Wet van 4 december 1906, betreffende het verlenen van concessies en vergunningen tot het gebruik van domein en voor de aanleg en de exploitatie van werken van openbaar nut als t.b.v. andere doeleinden
- Wet van 29 november 1915, tot vaststelling van een Politiestrafwet
- Wet van 27 december 1929 tot vaststelling van bepalingen omtrent het oprichten van inrichtingen, welke gevaar, schade of hinder kunnen veroorzaken
- Wet van 2 oktober 1931 tot uitvoering van het Regeringsreglement van Suriname, betreffende waterschappen
- Koninklijk besluit van 22 februari 1938 no. 41, houdende verbodsbepalingen met betrekking tot watervergaarplaatsen in Paramaribo en voorschriften inzake verplichte aansluiting aldaar aan de waterleiding
- Wet van 3 april 1954, houdende voorzieningen tot bescherming en behoud van de in Suriname aanwezige natuurmonumenten
- Wet van 7 december 1972, tot regeling van de handel in en het gebruik van bestrijdingsmiddelen
- Grondwet van 1975
- Decreet van 8 mei 1986, houdende algemene regelen omtrent de opsporing en ontginning van delfstoffen
- Wet van 29 juni 1989, houdende regels betreffende de inrichting en de bevoegdheden van de Regionale Organen
- Wet van 18 september 1992, houdende voorzieningen met betrekking tot het bosbeheer, alsmede de bosexploitatie en de primaire houtverwerkingssector
- Ontwerp wet van ...., houdende regels voor duurzaam milieumanagement (Milieuwet)
- Ontwerp wet houdende regels betreffende de bescherming van grondwaterwingebieden
- Ontwerp wet houdende regels betreffende het onttrekken van grondwater
- Ontwerp wet houdende regels betreffende het toezicht op de drinkwaterkwaliteit van waterbedrijven

CHAPTER 1
GENERAL INTRODUCTION

'There is enough freshwater in the world to meet the existing and future needs of the world's population.'[1]

1.1 PAINTING THE CONTEXT

Given the importance of water, it is a worrying fact that over 748 million people still do not have access to improved water and billions do not have access to safe drinking water.[2] This situation can rightfully be classified as a crisis[3] which has severe impacts on human health and the economy.[4] For example, a lack of safe drinking water makes people prone to waterborne diseases such as diarrhoea.[5] It also negatively affects economic activities and poverty reduction.[6] In sub-Saharan Africa the number of people without access to safe drinking water increased by 63 million between 1990 and 2011 as a result of population growth and poor infrastructure.[7] But also within the global water crisis a pattern of neglecting the needs of vulnerable groups in society can be identified across planning and resource allocation. These vulnerable groups can often be categorised along ethnic, geographical, and socio-economic divides but also along environmental lines.[8]

---

1 Eyal Benvenisti, 'Collective Action in the Utilization of Shared Freshwater: The Challenges of International Water Resources Law' (1996) 90 The American Journal of International Law 384. p. 384.

2 UN-Water, *The United Nations World Water Development Report 2015: Water for a Sustainable World* (UNESCO 2015). p. 19.

3 United Nations Development Programme, *Human Development Report 2006 Beyond Scarcity: Power, Poverty and the Global Water Crisis* (2006) <www.undp.org/content/dam/undp/library/corporate/HDR/2006 Global HDR/HDR-2006-Beyond scarcity-Power-poverty-and-the-global-water-crisis.pdf>. p. v.

4 (UN-Water, 2015) p. 93.

5 WHO/UNICEF, 'Drinking Water Equity, Safety and Sustainability. Thematic Report on Drinking Water' (2011). p. 41.

6 Naho Mirumachi, 'Water and Sustainable Development' in Michael Redclift (ed), *Routledge International Handbook of Sustainable Development* (Taylor and Francis 2015) <www.un.org/waterforlifedecade/waterandsustainabledevelopment2015/images/sustainable_development_eng.pdf>. p. 142.

7 UNICEF and WHO, 'Progress on Drinking Water and Sanitation 2013 – Update' (2013)., p. 36.

8 Catarina De Albuquerque, 'Report of the Special Rapporteur on the Human Right to Safe Drinking Water and Sanitation' (2013). p. 14.

In a nutshell, it can be said that the water crisis is 'both a natural and a human-made phenomenon. There is enough freshwater on the planet for seven billion people[9] but it is distributed unevenly and too much of it is wasted, polluted and unsustainably managed.'[10] This crisis has many factors which are interrelated and complicate matters when suitable solutions have to be found. The main factors contributing to the water crisis can be listed as follows:

The (mis)management of water

In its first World Water Development Report, the United Nations identified that '*the water crisis is essentially a crisis of governance and societies are facing a number of social, economic and political challenges on how to govern water more effectively.*'[11] Governance in this area is largely failing due to a lack of adequate water institutions and fragmentation in institutional arrangements.[12] For instance, failing to recognise that water is a topic which cuts across sectors means that many states approach water as a separate and independent sector within policy areas. At first sight, this may seem appropriate, but this approach results in a further compartmentalisation of water by dividing it between departments, policies and other authorities which then become responsible for developing separate strategies for health, food, agriculture and energy, to name just a few areas.[13] These strategies all apply to the same resource and, if not coordinated properly, could increase competition for water and increase water scarcity.

World population dynamics

The world population, totalling more than 7.5 billion in 2016, is expected to increase by almost one billion people within the next twelve years, according to the UN.[14] This population growth increases demand for and stress on all available resources including water supplies. In relation to this population growth, it is also necessary to know where these people will be living. Currently, water stress is highest in Africa and Asia and these are also the regions which are expected to see the greatest population growth in the years to come.[15] People go to where the water is and if those living in these regions cannot meet their water needs this may lead to displacement. In its first World Water Development Report, the UN confirmed that in 1998/99 as a result of the contamination and overuse of river basins, 25 million individuals had become environmental refugees.[16]

---

9 Nonetheless, the overall demand for water, when combining all the sectors together, exceeds the amount of water which is available. However, the global water availability is enough to meet the domestic and hygienic needs of everyone.

10 Www.un.org/waterforlifedecade/scarcity.shtml.

11 The World Water Development Report: "Water for People, Water for Life", Executive Summary p. 30.

12 John Scanlon, Angela Cassar and Noemi Nemes, *Water as a Human Right?* (IUCN 2004). p. 17.

13 UNESCO, *Managing Water under Uncertainty and Risk*, vol 1 (UNESCO 2012). p. 127.

14 UNFPA World Population Prospects, The 2012 Revision Key Findings and Advance Tables www.unfpa.org/webdav/site/global/shared/documents/news/2013/KEY%20FINDINGS%20WPP2012_FINAL-2.pdf.

15 Www.un.org/waterforlifedecade/scarcity.shtml.

16 Www.unwater.org/publications/publications-detail/en/c/210593/

Increasing water consumption

Water resources, especially groundwater, are being depleted at a high rate in order to meet the high water demands of different water users such as agriculture, industry and households.[17] Water for irrigation purposes has increased drastically[18] and, globally speaking, 15–35% of the withdrawals are estimated to be unsustainable.[19] In combination with demands from industry, freshwater withdrawals has increased sevenfold.[20] And the water demand for industry purposes is expected to rise with the rapid industrial development of developing states.

Water pollution

Another contributing factor to the water crisis is water pollution. Water pollution and decreasing water quality are caused by both natural and human agents in such a way as to render water unsuitable for several uses including for drinking.[21] Examples of natural pollutants are the high amount of arsenic contamination of groundwater in Bangladesh[22] and saltwater intrusion, which affects the quality of groundwater in coastal areas.[23] Furthermore, human activities which create the most environmental pollutants such as mining can cause environmental damage and pollute surface and groundwater.[24] Also, inadequately treated wastewater is a major polluter. A million tons of human waste have been, and still are, indiscriminately discharged into water bodies, causing water quality to decline.[25]

Environmental degradation

Even though environmental degradation is intertwined with water depletion and the pollution of water for human use, it is also necessary to pay attention to environmental

17 Ibid. p. 87.

18 Scanlon, Cassar and Nemes. p. 16.

19 World Business Council for Sustainable Development, 'Facts and Trends: Water' (2006). p. 5.

20 Scanlon, Cassar and Nemes. p. 16.

21 Catarina De Albuquerque and Virginia Roaf, 'Monitoring', *Realising the human rights to water and sanitation: A handbook by the UN Special Rapporteur Catarina De Albuquerque* (2014). p. 25; Mark D Sobsey, 'Managing Water in the Home: Accelerated Health Gains from Improved Water Supply' (2004) 8 World Health 1 <www.bvsde.paho.org/bvsacd/who/sobs.pdf>. p. 1.

22 Lorraine C Backer and others, 'Naturally Occurring Water Pollutants' in Janine MH Selendy (ed), *Water and Sanitation-Related Diseases and the Environment: Challenges, Interventions, and Preventive Measures,* (John Wiley & Sons, Inc 2011). p. 273.

23 Ibid. p. 282.

24 See Kerry L Hannon and Robert S Lawrence, 'Antropogenic Sources of Water Pollution: Parts 1 and 2' in Janine MH Selendy (ed), *Water and Sanitation-Related Diseases and the Environment: Challenges, Interventions, and Preventive Measures* (First, John Wiley & Sons, Inc 2011).

25 See Asit K Biswas, 'Water as a Human Right in the MENA Region: Challenges and Opportunities' (2007) 23 International Journal of Water Resources Development 209 <www.tandfonline.com/doi/abs/10.1080/07900620701523277> accessed 8 March 2013. p. 210. Worldwide wastewater has been discharged into the environment. During the 1960s the water quality declined and mostly industrialized countries started implementing waste water management to ensure the quality of water sources. Developing countries have not yet been able to implement such strategies.

concerns which often lose out in competition with human demands.[26] When water is reassigned and removed from the environment less water is available to protect and preserve rivers, streams, and wetlands, to restore aquatic and riparian habitats and to sustain threatened and endangered species as well as the water quality needed to support the ecosystems.[27] This factor adds to the loss of wetlands, the environmental wealth of the earth and threatens marine life but also affects the availability and suitability of water for human purposes in the long term.[28]

All of these factors contribute to the water crisis and are exacerbated by climate change and the economic crisis[29], which both disproportionately affect those which are already vulnerable when it comes to water.[30] The water crisis affects all stakeholders; nevertheless, those who lack access to water are the most negatively affected as they currently do not have access to safe water and the crisis might prevent them from gaining access in the future. This is often not an issue of absolute water shortages relating to physical supply.[31] For those who do not have access to water, only in rare cases is it caused by an absolute lack of water. What is more often the case is that the origin of the crisis for those lacking access lies in inequality, uneven power relations and faulty water management systems which aggravate the crisis.[32] When individuals do not have access to sufficient amounts of safe water, they depend on unimproved sources to meet their water needs such as creeks and rivers. If these sources are not protected and are highly susceptible to pollution, these individuals and groups run high health risks.[33]

The recognition of the human right to water provides an option, especially for vulnerable groups to base their claim on in order to have their access to water recognised as a human right and improved.[34] However, this approach should not be seen as a separate issue. The challenge of gaining access to water is situated within the current water context. A context in which water resources are limited due to pollution and depletion and are inequitably distributed. This study argues that states should ensure the sustainable management of water resources in order to support

---

26 See for instance Julia Ko and William F Donahue, *Moving Waters : Water Management Options to Achieve Social, Economic, and Environmental Goals* (Water Matters Society of Alberta 2012). p. 7.

27 Alezah Trigueros, 'The Human Right to Water : Will Its Fulfillment Contribute to Environmental Degradation ? The Human Right to Water : Will Its Fulfillment Contribute to Environmental' (2012) 19 Indiana Journal of Global Legal Studies 599. p. 602.

28 Patricia Avila-garcía, 'The Human Face of Water Security' <http://link.springer.com/10.1007/978-3-319-50161-1>. (Avila-García, 2017) p. 73.

29 Catarina De Albuquerque and Virginia Roaf, 'Principles', *Realising the human rights to water and sanitation: A handbook by the UN Special Rapporteur Catarina De Albuquerque* (2014). p. 76.

30 See Stephen Humphreys, *Human Rights and Climate Change* (Stephen Humphreys ed, Cambridge University Press 2009).

31 United Nations Development Programme.ibid. p. 133.

32 Ibid. p. 133.

33 Elliot Curry, 'Water Scarcity and the Recognition of the Human Right to Safe Freshwater' (2010) 9 Northwestern Journal of International Human Rights 103. p. 107.

34 Inga T Winkler, *The Human Right to Water* (2015).p. 8.

the realisation of the human right to water for those who are currently underserved and unserved. This study looks at obligations and requirements which stem from international human rights law, sustainable development law, international water and environmental law and policy to shape this sustainable management of water resources by states.

1.2 RESEARCH QUESTIONS

In order to meaningfully crystallise the human right to water, a broader approach needs to be taken as water is more than a social resource and plays a vital role across various sectors.[35] For the purpose of maintaining adequate access to safe water, it is necessary to maintain a safe supply of water, namely the overall river basin management, groundwater use and other works that are significant to meaningfully strengthen and uphold a human right to water.[36] In this light, interesting questions are those which concern the implications which a human right to water might have for water allocation in order to meet not only basic human needs (both present and future) but also the needs of the environment, especially the ecosystems providing these services. This study aims to explore this topic. It examines and assesses the potential of a human right to water in meeting both the basic needs of those who lack access while taking into account short and long-term environmental concerns. It does so by incorporating a sustainable approach as a way of finding a correct balance between the competing demands of humans and ecosystems. The theoretical conclusions concerning this sustainable approach are then explored in the context of case studies in order to test its applicability.

In order to structure the exploration of the above-mentioned, the following general research question has been developed:

How can the concept of *sustainable development* contribute to the sustainable realisation of the *human right to water* for *vulnerable groups*?

This question will be explored by addressing the following sub-questions:

1. Which groups and individuals should be considered 'vulnerable' in relation to the human right to water? And which entitlements and obligations accompany the qualification of 'vulnerable'?
2. How did the human right to water develop within international law and what does it entail? What are the legal foundations of this right and which obligations does it create for states?

---

35 Inga Winkler, 'Policy Guidance Workshop: The Human Right to Water and Sanitation "Translating Theory into Practice"' (2008). p. 5.

36 Scanlon, Cassar and Nemes. p.27.

3. What does sustainable development mean in relation to water resources? Which principles from international environmental law can inform water resources management in order to support the realisation of the human right to water?
4. Which aspects of sustainable development can enhance the realisation of the human right to water for vulnerable groups and in which manner?
5. How can a sustainable human right to water be assessed within the local context?
6. How can a sustainable human right to water for vulnerable groups be fulfilled in Suriname?

The study then looks at how this bridging of the two fields of human right and sustainable development could contribute to the realisation of the human right to water for vulnerable groups and individuals at the national level. For this study, Suriname was chosen to serve as an illustration. This state provides an interesting illustration as it has a great wealth of fresh water resources, despite this universal access to water has not been reached as nationally between 72.6 – 79.4 percent of the population has access to safe water. There is great disparity in access between areas.[37] In the urban areas majority of the population, between 88 – 98 percent, use drinking water that is piped into their house or into their yard. However, in the rural interior great proportions rely on rainwater collected in unsafe manners as drinking water and river or creek water for household purposes. Only between 20 – 40 percent of the population has access to safe drinking water. In addition, Suriname is confronted with issues of pollution as a result of economic activities. Suriname therefore serves as an appropriate example to illustrate the connections between sustainable development and the human right to water for vulnerable groups. Besides these reasons, accessibility to information and knowledge about the legal system, language and socio-economic situation in Suriname by the author motivated the selection of Suriname for the case-studies.

In the following sections, the concepts which are central to this research – the human right to water and sustainability in relation to water – are introduced. After this the theoretical framework (1.4) will be outlined and an overview of the methodology (1.5) is given, which is followed by an introduction to the case studies (1.6) that will be undertaken. Subsequently, the significance of the research (1.7) is discussed. Finally, the scope and structure of the research are provided in section 1.8.

1.3 INTRODUCTION: CENTRAL CONCEPTS

The human right to water is a curious right within international human rights law, as it is not explicitly mentioned in the three most authoritative documents outlining

[37] See Government of Suriname and United Nations Country Team Suriname, 'Suriname MDG Baseline Report' (2005). p.58 and UNICEF, 'Suriname Multiple Indicator Cluster Survey 2010' (2013). p. 72.

fundamental human rights.[38] Neither the Universal Declaration of Human Rights (UNDHR), the International Covenant on Economic, Social and Cultural Rights (ICESCR), nor the International Covenant on Civil and Political Rights (ICCPR), which establish an authoritative list of human rights, explicitly recognise a human right to water. In the drafting of these instruments, especially the ICESCR, the inclusion of water seems to have been overlooked as at that time water was seen as being widely available.[39] Nevertheless, the recognition of access to water as a basic human need emerged from the international environmental field around the mid-1970s before it crossed over to the human rights field. The human right to water has been made the subject of debate by academic scholars and international organisations, who have approached this problem in several ways. And it has indeed been recognised by some states in different forms and in national instruments,[40], while others question whether such a right even exists.[41] This study provides an extensive and critical examination of the current status and content of this right.

A human right to water can be an important key in addressing access to water for the currently underserved and unserved. Having access to water framed as a human right therefore creates several benefits which will be discussed in this book. One of the most important benefits, however, is that access to water is seen as a claimable right, establishing duties for states and not being seen as merely a basic human need. Individuals can use the right to hold states accountable and do not have to depend on the willingness of governments alone. Human rights have a legally binding status which provides legitimacy and authority[42] and which can form an incentive for the prioritisation of this right within national laws and policies.[43] For these reasons, human rights can provide for a framework which can be compelling and comprehensive and can provide a basis for individual claims to water.

In the global water crisis where water resources are at risk of depletion by demand exceeding the available water supply and where water is increasingly polluted, sustainability is a very important requirement, as these occurrences are bound to increase.[44] For this reason a connection between the human right to water and sustainability should be fostered. The concept of sustainability is no stranger in relation

---

38 Takele Soboka Bulto, 'The Emergence of the Human Right to Water in International Human Rights Law : Invention or Discovery?' (2011) 7. p. 2.

39 Stephen C McCaffrey, 'A Human Right to Water : Domestic and International Implication' (1992) 5 Georgetown International Environmental Law Review 1.

40 See Thorsten Kiefer and others, *Legal Resources for the Right to Water and Sanitation* (2nd edn, Centre on Housing Rights and Evictions 2008) <www.cohre.org/water> for examples of states which have included a constitutional provision for human right to water.

41 Stephen Tully, 'A Human Right to Access Water? A Critique of General Comment No. 15' (2005) 23 Netherlands Quarterly of Human Rights 35. p. 42.

42 Manisuli Ssenyonjo, *Economic, Social and Cultural Rights in International Law* (2nd edn, Hart Publishing 2016). p. 13.

43 Inga Winkler, *The Human Right to Water Significance, Legal Status and Implications for Water Allocation* (First, Hart Publishing 2012).p. 150.

44 Joseph W Dellapenna, 'Is Sustainable Development a Serviceable Legal Standard in the Management of Water?' [2004] Water Recources Update 87. p. 89.

to water.[45] It can even be seen as an emerging and still developing legal requirement.[46] Sustainability is part of both hard law and soft law instruments and emerges in international law in various ways. Nevertheless, its content and how it is to be applied are part of the ongoing debate.[47]Terms such as sustainable development and sustainable utilisation are all expressions of sustainability and are often used interchangeably even though they differ in their meaning.[48] This study focusses on sustainable development which has been formulated in the Brundtland Report as: 'development that meets the needs of the present without compromising the ability of future generations to meet their own needs.'[49] It is the objective of this study to explore how sustainable development in relation to water resources and its complementary principles can be used to secure access to water for vulnerable groups in a meaningful manner.

Vulnerable groups are the focus of this study. As mentioned 748 million people lack access to safe drinking water. The needs of vulnerable groups are often disregarded in allocating water.[50] And with both the quantity and quality of this important resource depleting, the amount of people lacking access to water could increase if interventions do not take place. From a human rights perspective, vulnerable groups require special attention.[51] This study explores which groups are entitled to this special attention and how states can meet their obligations. Vulnerable groups as such serve as a 'litmus test' for the overall situation of the human right to water. As it can be assumed that if the circumstances in a state are such that even vulnerable groups have access to water as they are the 'high hanging fruits', it can be assumed that all others have as well.[52]

1.4 THEORETICAL FRAMEWORK

This chapter started by emphasising the global water crisis in which especially the interests and needs of ecosystems and vulnerable groups are losing out. It is the purpose

45 For instance, the Sustainable Development Goals, an important soft law instrument, include goal number 7: 'to ensure environmental sustainability', with as one of its targets being to 'halve, by 2015, the proportion of the population without sustainable access to safe drinking water and basic sanitation'.

46 Ibid. p. 88. See on the debate on development of resources including water (Boyle and Freestone, 1999)

47 See Peter Orebech and Martin Chanock, 'Towards Sustainability : The Basis in International Law' in Fred Bosselman and others (eds), *The Role of Customary Law in Sustainable Development* (Cambridge University Press 2006).

48 (Dellapenna, 2004) p. 89.

49 Brundtland Commission, 'Our Common Future: Report of the World Commission on Environment and Development' (1987). Chapter 2.

50 See chapter 3.

51 Audrey R Chapman and Benjamin Carbonetti, 'Human Rights Protections for Vulnerable and Disadvantaged Groups: The Contributions of the UN Committee on Economic, Social and Cultural Rights' (2011) 33 Human Rights Quarterly 682 <http://muse.jhu.edu/content/crossref/journals/human_rights_quarterly/v033/33.3.chapman.html> accessed 17 December 2013. p. 726.

52 Catarina De Albuquerque, 'Report of the Independent Expert on the Issue of Human Rights Obligations Related to Access to Safe Drinking Water and Sanitation' (2010). p. 14.

of the following section to lay out the framework which will be used to analyse the current situation and to provide possible strategies and tools. As mentioned, this research takes a human rights perspective concerning the global water crisis and explores the incorporation of sustainability in the human right to water in order to secure access to water for vulnerable groups for domestic and hygienic uses. For this purpose two distinct perspectives are combined, namely sustainability in the context of water and a human rights-based approach to water. How these two are combined within the framework and which definitions are used will be explained in the following subsection. Firstly, it is explained why this approach has been selected in this form, why these two perspectives are deemed relevant for vulnerable groups and, briefly, how the case studies fit in.

Particularly vulnerable groups, which are the focus of this study, can profit from the human right to water combined with the inclusion of sustainability as they suffer disproportionately from violations of their right to water[53] and need appropriate measures for redress. These groups are often neglected in resource allocation and they are unable to protect themselves effectively against environmental degradation.[54] They often live in highly polluted areas under circumstances which prevent or hinder them in saying: not in my backyard![55] Two major issues with which these groups are confronted are the small amounts of water they have access to and water being polluted making it unsuitable for domestic uses.[56] These issues are related to the aims of sustainability which are to preserve water and to have it available for human usage as well as to protect ecosystems.

In addition, realising access to safe and sufficient water for this group not only requires water resources to be clean and sufficient in quantity but also requires a process to support the realisation of these requirements. For instance, available water resources can be free from pollution but specific groups can be prohibited from gaining access on the basis of their stigmatisation or marginalisation. For example, persons with disabilities are often stigmatised and discriminated against within many countries. Making use of public services or even household facilities can be negatively influenced by these perceptions. 'Fears that persons with disabilities would contaminate water sources or that they would make the latrines dirty are frequently reported.'[57] In this

53 Statement by the Special Rapporteur on the human right to safe drinking water and sanitation during the discussion at the Permanent Forum on Indigenous Issues, Tenth Session, 11th meeting 24 May 2011 available at: www.un.org/News/Press/docs/2011/hr5061.doc.htm (last retrieved on 7 March, 2014) 'Asking who exactly did not have access and why, Catarina de Albuquerque said it was always the same people – the marginalized, the poor and those without a political voice.'

54 Jon Barnett, 'Human Rights and Vulnerability to Climate Change' in Stephen Humphreys (ed), *Human Rights and Climate Change* (Cambridge University Press 2010). p. 266 and see (Eakin, H. and A. Luers 2006)

55 Oxfam (2012) available at www.oxfam.org/sites/www.oxfam.org/files/bp157-left-behind-by-the-g20-190112-en.pdf (last retrieved: 7 March, 2014)

56 United Nations Development Programme. p. 75–109.

57 N Groce and others, 'Water and Sanitation Issues for Persons with Disabilities in Low- and Middle-Income Countries: A Literature Review and Discussion of Implications for Global Health and International Development.' (2011) 9 Journal of water and health 617 <www.ncbi.nlm.nih.gov/pubmed/22048421> accessed 7 April 2014. p. 619.

case substantive requirements are met but in practice access is prevented due to social issues and that is where the human right to water steps in as an additional perspective alongside sustainability.

The human right to water provides these groups with procedural rights such as the right to participation and information which will allow their interests to become visible. This approach may also assist in identifying priorities as a way of dealing with competing interests and demands.[58] These elements are highly relevant for the realisation of the right to water for vulnerable groups and provide a comprehensive basis from which to look at sustainability.

The human rights framework can provide vulnerable groups with a holistic approach to dealing with issues concerning water.[59] The purpose of human rights is to accommodate important claims concerning basic needs and to provide redress for those who are affected.[60] Human rights and the international bodies overseeing their implementation give special attention to vulnerable groups.[61] The special mechanisms and rights given to them in this approach can provide them with additional support. Therefore, assessing possible remedies at the international level is highly relevant for them, especially if national legislation to protect their access to this fundamental resource is not available or if such legislation is not backed up by enforcement. Without an international approach, which is universally applicable to protect the essential human need to water, 'the right would most likely not be protected precisely where it is needed most: those to whom the right is most crucial, would be those who lack it.'[62]

In addition, if international human rights standards are integrated within the national legal framework this could be an appropriate response to challenges experienced by vulnerable groups.[63] When using the human rights approach to support their claims, obstructions caused by other legal domains can be identified. Water management is often fragmented at the national level. As access to water cuts across disciplines, it can reform or influence relevant laws by allowing for the interpretation of human rights values where the existing legal framework allows for the appropriation and pollution of water without considering the negative impacts on humans and on nature.[64]

---

58 See Winkler, *The Human Right to Water Significance, Legal Status and Implications for Water Allocation*. on the human rights implications for water allocation and prioritization.

59 Chapman and Carbonetti. p. 682.

60 Nandita Singh and others, 'Accessing Water through a Rights-Based Approach: Problems and Prospects Regarding Children' (2012) 14 Water Policy 298 <www.iwaponline.com/wp/01402/wp014020298.htm> accessed 11 March 2013. p. 303.

61 See Chapman and Carbonetti.

62 Pierre Thielbörger, 'Philosophical and Conceptual Approaches to a Human Right to Water', *The Right(s) to Water* (Springer Berlin Heidelberg 2014) <http://link.springer.com/10.1007/978-3-642-33908-0> accessed 12 March 2014. p. 125.

63 Water Governance Facility, 'Human Rights-Based Approaches and Managing Water Resources: Exploring the Potential for Enhancing Development Outcomes' (2012). p. 8.

64 Hugo Tremblay, 'A Clash of Paradigms in the Water Sector? Tensions and Synergies Between Integrated Water Resources Management and the Human Rights-Approach to Development' (2011) 51 Natural Resources Journal 307. p. 321–323.

For example, in India the right to life, as formulated in the Constitution[65], is interpreted in a manner which covers both the right to water and the right to live in a healthy environment. This provision formed the basis of complaints against Coca Cola concerning groundwater depletion and pollution and allowed adjustments to be made to contracts between the government and Coca Cola.[66] This example shows that this approach can make a significant contribution to achieving sustainability in managing the water resource basis of the human right to water. Also, considering the potential influence of the HRBA at the policy making level, the authorities might be motivated to amend permit regulations by requiring agents to take domestic use and environmental concerns into account.[67]

Building on the importance of tackling both substantive issues relating to water quality and quantity, as well as issues stemming from the societal sphere, such as a lack of participation in water management, the human right to water will be the starting point of the analysis within this study and it will look at the contribution that sustainability could make to the human right to water and not the other way around. This is because addressing the issue the other way around, taking sustainability as the starting point and incorporating the human rights perspective into it, would be less appropriate for this group in particular. Firstly, sustainability does not narrow down its focus to providing water for drinking and domestic purposes, which is the major challenge faced by vulnerable groups. Secondly, it does not provide mechanisms which allow for redress, an element which can be very useful for vulnerable groups.[68] Nevertheless, the concept of sustainability can contribute valuable aspects which could further assist in developing the human right to water. Several of these aspects are introduced in the following subsection.

1.5 OVERVIEW OF THE METHODOLOGY[69]

This research aims not only to present recommendations concerning the theoretical side of this issue, but also to analyse its possible implications for the field and therefore includes a case study. The following section provides an overview of the methods used in conducting the research presented. It explains which methods and sources were

65 Article 21: 'no person shall be deprived of his life or personal liberty except according to procedure established by law'. The first case which recognized the right to live in a healthy environment as being a part of Article 21 was Rural Litigation and Entitlement Kendra vs. State of U.P., AIR 1988 SC 2187 (Popularly known as the Dehradun Quarrying Case).

66 The national case of India – Perumatty Grama Panchayat vs State of Kerala (2003.12.16) (Coca-Cola groundwater exploitation case) available at https://www.elaw.org/node/1410 (last retrieved on 7 April 2014).

67 Ibid. p. 345.

68 Water Governance Facility. p. 12.

69 These methods are generally used in this study. Chapter 7 includes an additional overview on methodology explaining the research methods used for that particular chapter.

used to develop the theoretical conclusions and the context-specific recommendations relating to the case study.

This book involves a combination of two different research methods, namely desk-based legal research using legislation, academic literature and other relevant sources such as reports. Secondly, this book uses qualitative legal empirical research involving non-legal sources because it includes information gathered from interviews and non-legal document analysis.[70] This methodological pluralist approach is applied for several reasons. Firstly, to enlarge the understanding of the research problem since multiple approaches will shine light on different aspects of the problem, thereby exposing the bottlenecks. Secondly, this method also has an exploratory function as it has assisted in examining whether certain issues which seemed problematic in theory also existed in practice. And thirdly, using two different methods allows for the results to be verified.[71]

For instance, sustainability might be a goal which is specified on paper but it might not correspond with the values that are present in the society where it is supposed to be implemented and may therefore experience difficulties in its application. The insights which both methods bring to the table will inform the framework which is presented in Chapter 5 and the circumstances and requirements necessary for the enjoyment of a sustainable human right to water by vulnerable groups.

1.5.1 DESK-BASED RESEARCH

The theoretical part of this research was carried out by employing desk-based research, meaning that the concepts of vulnerability, the human right to water and sustainability in the context of water were analysed by means of desk-based research. This desk-based research includes an analysis of conventional legal sources and literature.[72] The examination of conventional sources of law concerned (national) legislation, treaties, general principles, case law, doctrine and customary law.[73] In addition, the relevant academic legal literature was studied as an aid in interpreting sources of law and for background information.

The relevant literature was found through appropriate combinations of search terms such as the human right to water/ access to water/ the right to health/ vulnerable

---

70 Lisa Webley, 'Qualitative Approached to Emperical Legal Research' in Peter Cane and Herbert Kritzer (eds), *The Oxford Handbook of Empirical Legal Research* (Oxford University Press 2010). p. 926.

71 Ibid. p. 11.
Tarrow argues, "Triangulation is particularly appropriate in cases in which quantitative data are partial and qualitative investigation is obstructed by political conditions. Margaret L Satterthwaite and Justin C Simeone, 'A Conceptual Roadmap for Social Science Methods in Human Rights Fact-Finding' in Philip Alston and Sarah Knuckey (eds), *The Transformation of Human Rights Fact-Finding* (Oxford University Press 2016).

72 The classic methodology applied in legal science consists of an analysis of conventional sources of law, meaning: legislation, general principles of law, case law and doctrine by means of a desk-based study (see Ellen Desmet, *Indigenous Rights Entwined with Nature Conservation* (Intersentia 2011). p. 36).

73 Webley.

groups/ sustainability/ sustainable development/ environmental law/ water law. This was used to identify and substantiate principal sources and discussions. Articles and other literature were found through Google Scholar and different databases for journals and books by institutions such as Utrecht University, Leiden University and the Peace Palace Library. Searching these databases was supplemented by the 'snowball method.' These search methods were applied to the discussion of the three main concepts of this study: vulnerable groups (Chapter 2), the human right to water (Chapter 3) and sustainability (Chapter 4).

The concept of vulnerability as understood by the UN Committee on Economic, Social and Cultural Rights is analysed and evaluated from the point of view of the environmental issues concerned, but on a very limited basis. This analysis therefore requires the present understanding of vulnerability to be identified within international human rights law. The study limits itself to the concept within the international human rights arena because the human right to water as understood at that level forms the core of this study. Documents such as general comments, resolutions, state reports and concluding comments are considered in order to draw conclusions on the definition, criteria and rights assigned to vulnerable groups within international human rights law. In order to broaden the concept of vulnerability, ideas are used from other related fields such as environmental justice.

The concept of sustainability within international law also required analysis through desk research. The legal fields covered for this analysis are international law, environmental law and international water law. The sources examined include: customary international law, treaties, agreements and other documents of UN bodies (such as the Sustainable Development Goals), policy documents, declarations and resolutions resulting from conferences, and also documents by NGOs (such as the ILA and IUCN). These instruments and documents allow the notion of sustainability to be explored and to relate it to water. These documents are not only legal documents but also documents related to international policy. In addition, the legal documents differ in their legal status.

As mentioned, both the UN General Assembly and the Human Rights Council have recognised the human right to water by stating so in several resolutions. This recognition has political and legal significance as part of state practice.[74] Nevertheless the existence of a human right to water needs evaluation from a legal perspective. This exploration requires a detailed analysis of the legal foundations of a human right to water. General principles, human rights treaties and customary international law are considered. General principles are found in the decisions of international and national courts[75], such as the decisions of the International Court of Justice. These sources need to show consistency in recognising such a right. Human rights treaties are of course discussed,

74 Thorsten Kiefer and others, *The Human Right to Safe Drinking Water and Sanitation in Law and Policy – a Sourcebook* (WASH United, Freshwater Action Network, Waterlex 2012). p. 48.

75 Infact, Article 38 includes judicial decisions (by both international and municipal tribunals) and scholarly writings on "subsidiary means for the determination of rules of law;" in other words, these are not authorities, they are rather evidence of the sources of international law.

beginning with the most authoritative human rights treaties at the global level, the ICESCR and the ICCPR and regional instruments in as far as they are relevant for the case studies. Following this analysis other relevant treaties will be discussed such as the CEDAW and the CRC. Aside from these documents, other relevant texts from UN bodies such as general comments from human rights bodies and other declarations, resolutions and statements by states, UN bodies and NGOs are relevant to the exploration of a possible human right to water. These instruments are considered to be instruments of 'soft law', but they nevertheless have a considerable impact. Soft law can assist by filling in gaps, guiding interpretations and signalling the emergence of international law.[76]

1.5.2 QUALITATIVE LEGAL EMPIRICAL RESEARCH

Besides the traditional legal sources also non-legal sources, such as reports and non-legal literature, are used to explain the relevant contexts, such as the physical and social context in which the human right to water needs to function. Water is by its nature highly intertwined with the everyday lives of people and therefore studying access to water is particularly contextual. It is in the national context where local processes, strategies and mechanisms are at play and which are instrumental for the implementation of international norms such as sustainability and the human right to water. Therefore it is necessary not only to take notice of legal issues through literature and legal sources but also to look at how these concepts and issues function in 'real life' by means of case studies. Looking at the workings of the concepts in practice requires qualitative legal empirical research.

The qualitative research[77] conducted in this study has been collected through two main methods, namely: interviews and an analysis of the relevant documents. More specifically, the case studies form a complementary part of the desk-based research (providing the theoretical background) and combine a range of data collected from interviews with stakeholders and experts[78] and an analysis of policies and academic (non-legal) literature.[79]

Within this research, the case study has both a descriptive and an explanatory function. The case study has been designed in such a manner that it illustrates the situation as it is and determines why and how vulnerable groups are unable to enjoy a sustainable human right to water in their respective context. Ultimately the aim is to provide recommendations to improve this situation. Experts and relevant stakeholders

---

76 Particia Birnie, Alan Boyle and Catherine Redgwell, *International Law and the Environment* (Third, Oxford University Press 2009).

77 It is difficult to provide a precise or widely accepted definition of qualitative research and the theory underpinning it because so much of the terrain is contested. Most researchers who conduct qualitative research would agree that it is socially concerned, examines phenomena in their social settings (if fieldwork is being undertaken) and considers those phenomena in context.

78 Consult appendices for interview protocol and list of consulted experts and stakeholders. research.

79 Webley. p. 939.

play a crucial role in this part of the research. They provided knowledgeable information and legitimate opinions as well as represented different perspectives. Also through interviewing these stakeholders and experts, access through other sources was made possible. Some information was for instance inaccessible because of the distance involved or because not all relevant sources had been digitised.

The experts and stakeholders were asked to give their view on aspects which are relevant for realising the human right to water for vulnerable groups and how sustainability goals can be attained within the context of this right. Those who participated in the interviews are divided into three categories:

1. Policy and law makers, representing the viewpoint of authorities, both national and local, which apply and create law and policy. The value of their contributions rests on their expertise concerning the application, origins and intentions behind the relevant laws and policies.
2. Civil society experts and members of communities located in the rural interior providing insight into the impact of law and policy on vulnerable groups, their needs, and how these needs might be met.
3. Academic scholars, providing expert opinions about legal aspects and sustainability issues.

1.6 INTRODUCTION TO THE CASE STUDY

As explained, the water crisis is not a mere physical issue, so the availability of water, the use of water and the management framework do not properly correspond with each other.[80] This book partly explores the possible contributing factors in addressing the water crisis from a human rights perspective complemented by the concept of sustainability. The case study provides an opportunity to examine these issues as well as the main question: how can the concept of sustainability be incorporated into the human right to water in such a way as to enable the enjoyment of this right by vulnerable groups in the context of Suriname?

Undertaking the case study will show the possibilities and difficulties of putting the concept of sustainability and the human right to water into practice and aims to bring to light which factors in the legal and physical framework facilitate and prohibit access to water for vulnerable groups. The study aims to set out the requirements but also looks at how those requirements might work in practice. In this manner theory and practice can feed back into each other.

Suriname was selected as one of the two case study countries, representing a country with sufficient water resources available. At the beginning of this study Yemen was

80 Carlos Fernández-Jáuregi and Alberto Crespo Milliet, 'Water, a Unique Resource' in Mikel Mancisidor and Natalie Uribe (eds), *The Human Right to Water: Current Situation and Future Challenges* (UNESCO 2008). p. 29.

selected as the second case study, representing a country experiencing water scarcity. However, due to the current political unrest and armed conflict in that country, it was decided not to include Yemen as a research subject in a separate chapter. This is because not all collected information could be verified at the end of the study. The results which have been verified have been added throughout the book as examples.

The analysis of the case study is done by using an assessment framework which will be referred to as the 'Ten building blocks for vulnerable groups' sustainable access to water for domestic purposes' and was developed from a multi-disciplinary assessment method for water governance developed by Van Rijswick, Edelenbos, Hellegers, Kok, and Kuks (2014). This method has been expanded upon and systematically assesses relevant aspects, so-called building blocks, for the implementation of a sustainable human right to water for vulnerable groups. This framework includes the different disciplines of law, economics, and public administration and provides for knowledge on basic water systems. These elements are needed in order to provide context-specific recommendations.

Exploring the workings and understandings of the combined perspectives at the national and local level assists in resolving the bottlenecks and provides feedback for law and policy development. Tensions and imperfections need to be resolved and doing so at the local level on a case-by-case basis is the most appropriate method.[81] The case study offers the possibility for just this by looking at the national level in a specific physical, social and legal context and by assisting in resolving ambiguities. The national level can provide answers as to how and which decisions have to be made in the case of conflict.

Every context needs and deserves specific solutions for access to water to be realised equitably and sustainably. What might work in one context may not work in another and vice versa. This awareness of context-specific solutions is also accepted in international law as '[e]very State party has a margin of discretion in assessing which measures are most suitable to meet its specific circumstances.'[82]

The case study was selected[83] to provide insight into and to explore and show how the balancing of the two perspectives might work in practice.[84] The assessment framework developed can be applied to other national contexts. Access to water is such a complex issue and it cannot be explored properly when it is disconnected from the physical context. Looking at multiple and different settings will increase the probability that tools will be developed which have practical application and thereby increase the transferability of these tools and the designed framework.[85]

---

81 Tremblay. p.309.

82 (Committee on Economic, Social and Cultural Rights, 2002) para. 56.

83 The following criteria have been applied:
- Diversity in water-related issues: quality, quantity, affordability etc.
- Availability and accessibility of information.
- Location in the global south
- Presence of vulnerable groups that are confronted with freshwater issues.

84 See Miriam Van Bardeleben, *Implementing Sustainability* (2011).

85 Barbara Cosens and others, 'Identifying Legal, Ecological and Governance Obstacles, and Opportunities for Adapting to Climate Change' (2014) 6 Sustainability 2338 <www.mdpi.com/2071-1050/6/4/2338/> accessed 25 April 2014. p. 2341.

1.6.1 SURINAME

Suriname, the smallest mainland country in South America, has a great wealth of freshwater resources, referred to as 'white gold.'[86] Nevertheless, this country has to deal with two major issues. Firstly, providing access to safe water for domestic purposes for all those within its territory. Currently, the highest percentage of the population with access is in the urban areas reaching between 85 – 90 percent[87], followed by the rural area with 66.6 percent and, lastly, the Interior with just 20 percent.[88] This means that the national percentage of the population with access to safe drinking water is around 72.6 percent and shows that even though there is enough freshwater within the Surinamese territory, not everyone is benefiting from it.[89]

The second issue in relation to water that Suriname has to deal with lies in protecting its water resources from pollution and depletion while pursuing development. The Government considers water to be essential for its economic development and therefore plans to use it mostly for this purpose. Waste from human consumption, mining and agricultural activities are the biggest polluters and are hardly regulated. Legal provisions regulating water exploitation are also lacking.

Even though Suriname is not experiencing a water crisis which is as extreme as states in the Middle East or Sub-Saharan Africa, it is relevant to analyse and elaborate suggestions for improvement as Suriname is currently lagging behind in its obligation to realise the human right to water and it will jeopardise its water wealth if it continues with business as usual. The situation of Suriname shows that even though there is enough water, there are still restricting factors preventing vulnerable groups from gaining access. Suriname clearly demonstrates that water crises are not caused by water scarcity alone, but are the result of an interplay between many different factors. The importance of governance should also not be underestimated.

1.7 SIGNIFICANCE OF THE RESEARCH

This study seeks to answer the research questions mentioned through an analysis of the water situation of vulnerable groups and their place within international law, delineating the concept of sustainability in the context of water and exploring the (potential) legal foundations of the human right to water, state obligations and their normative content. Subsequently, this study examines the implications that sustainability has for the human right to water and how these implications affect the

86 Suriname is currently number 6 in the Water Poverty Index. 'The Water Poverty Index is a new holistic water management tool that is mainly relevant at the community level. It can be used to determine priorities for action and to monitor progress towards targets.'

87 NIMOS, 'Environmental and Social Analysis/environmental and Social Management Framework' (2010). p. 14.

88 Government of Suriname and United Nations Country Team Suriname. p. 42.

89 Ibid. p. 42.

managing of water sources used to realise the right. The study also looks at the national context of Suriname. The focus of this study is directed towards the state as the primary duty bearer and vulnerable groups as rights holders.

It should be noted that this study does not claim that the human rights-based approach and the sustainability approach are the only approaches which can be used to realise a sustainable human right to water for vulnerable groups.[90] It has chosen this path because, as will be shown, the developments surrounding the human right to water seem to head in that direction. This study therefore contributes to the academic discourse by clarifying how these two frameworks can be aligned for the benefit of the human right to water. What should be kept in mind is that a human right to water has to function within the context of hydrology, economics, geography, law and politics.[91] This human right nevertheless provides a legal strategy which could prove useful under certain circumstances which are explored in this research and contribute to solving a major social issue of current times.

1.8 SCOPE AND OUTLINE OF THE RESEARCH

The study is divided into eight chapters. Following this introduction, chapter two discusses who can be considered vulnerable in relation to water access. Vulnerable people are of concern as within practically every society certain individuals and groups structurally lack the enjoyment of human rights, especially economic, social and cultural rights.[92] In times of economic scarcity and of environmental degradation, the need to monitor the situation of vulnerable groups becomes even more urgent as they are disproportionately affected by these issues.[93] Analysing the situation of these groups serves as a litmus test for the realisation of the human right to water within a state. In addition, aiming for the realisation of this right for current vulnerable groups will set up the groundwork needed for the enjoyment of the right for future generations. Chapter 2 explores how vulnerability is conceptualised by the UN Committee on Economic, Social and Cultural Rights (UNCESCR), which has authoritatively interpreted the human right to water and argues that it does not provide sufficient protection for those vulnerable in relation to this right as it only concerns environmental issues on a very limited basis. It argues how vulnerability should be broadened in order to cover all

---

90 Winkler, 'Policy Guidance Workshop: The Human Right to Water and Sanitation "Translating Theory into Practice"'. pp. 14 – 15.

91 David B Brooks, 'Human Rights to Water in North Africa and the Middle East: What Is New and What Is Not; What Is Important and What Is Not' (2007) 23 International Journal of Water Resources Development 227 <www.informaworld.com/openurl?genre=article&doi=10.1080/07900620601097075&magic=crossref> accessed 8 March 2013, p. 227. This context has set boundaries for the answers found in this study.

92 Chapman and Carbonetti. p. 683.

93 Ibid. p. 693.

of vulnerable individuals are often disregarded in water allocation.[113] From a human rights perspective, however, vulnerable groups require additional attention to ensure that they are able to exercise their human rights, also their right to water. One of the purposes of the human rights framework is to achieve substantive equality.[114] In order to attain this, states are obligated under international human rights law to prioritise access for individuals and groups which are particularly vulnerable, even in times of austerity.[115] States are required to adopt targeted (temporary) positive measures and to redress, for instance, discrimination which could be the reason behind vulnerability.[116] An example of this can be found in relation to CEDAW's Article 4[117]: however, the adoption of temporary special measures for the purpose of addressing the structural and systematic discrimination of other vulnerable groups has also been recognised in relation to economic, social and cultural rights. CESCR General Comment No. 20 for instance states:

> *States parties must adopt an active approach to eliminating systemic discrimination and segregation in practice. Tackling such discrimination will usually require a comprehensive approach with a range of laws, policies and programmes, including temporary special measures.*[118]

Vulnerability plays an important role in the human rights framework. However, there is still quite some ambiguity about its meaning.[119] The term vulnerability is used in this research and it has been adjusted to fit the overall topic of this research by expanding on the views of the CESCR. This chapter discusses the following research questions: How is vulnerability framed by the CESCR and how should vulnerability be conceptualised to ascertain special protection for vulnerable groups in relation to the human right

---

Sources WHO Drinking-Water Quality Series Protecting Groundwater for Health MANAGING THE QUAALITY OF DRINKING-WATER SOURCES (2006).

113 OHCHR, *Fact Sheet No. 35, The Right to Water* (2010) <www.ohchr.org/Documents/Publications/FactSheet35en.pdf>. p. 14.

114 De Albuquerque and Roaf, 'Principles'. p. 14.
'In order to reach substantive equality, socially and culturally entrenched root causes must be addressed, discriminatory laws and practices repealed, and positive measures employed to achieve the equal enjoyment of rights.' UN-Water, 'Eliminating Discrimination and Inequalities in Access to Water and Sanitation' (2015). p. 48.

115 Chapman and Carbonetti. p. 693.
'Economic stress reinforces the physical and environmental stress due to water scarcity, pollution and hazards.' Úrsula Oswald Spring and Hans Günter Brauch, 'Securitizing Water' (2009) 4 Facing Global Environmental Change 175 <http://dx.doi.org/10.1007/978-3-540-68488-6_11>. p. 186.

116 Ssenyonjo. p. 143.

117 See CEDAW Committee, 'General Recommendation No. 25 on Article 4, Paragraph 1, of the Convention on the Elimination of All Forms of Discrimination against Women, on Temporary Special Measures' 10. and Ssenyonjo. p. 143.

118 CESCR General Comment. No. 20 paragraph 39. Another example can be found in CESCR General Comment No. 5 paragraph 18.

119 Ibid. p. 22.

to water? Which conditions make certain groups and individuals vulnerable? Which requirements does the designation of 'vulnerability' create for states?

This chapter answers these questions by firstly looking into the Committee on Economic, Social and Cultural Rights' views on vulnerability as it has formulated the most authoritative articulation of the human right to water. The Committee does not provide a clear definition of this term. Nevertheless, certain characteristics can be discerned from its practice. In addition, the views of the Committee will be expanded upon in order to develop a definition of vulnerability which is suitable for this research. Once the definition of vulnerability has been given, the chapter will proceed by outlining the obligations that accompany the qualification. This will be followed by a discussion of the categories of vulnerable groups for which rights to water have been explicitly formulated. The chapter will end with a conclusion.

2.2 THE LACK OF A CLEAR DEFINITION OF VULNERABLE GROUPS IN HUMAN RIGHTS LAW

The human rights framework refers to rights which apply to everyone, everywhere. These rights are often formalised in international conventions and protected by law.[120] Since the first developments formalising universal human rights in the Universal Declaration of Human Rights, different ideas have been propounded relating to recognising the vulnerabilities of individuals and allowing room for other particularities when implementing rights such as regional differences.[121] The recognition of vulnerability is meant to effectively deal with obstacles for individuals and to take into account the specific contexts in which individuals live in order to realise a minimum standard of living.[122]

Through its Committees the international human rights framework increasingly recognises that specific individuals might experience disproportionate difficulties in realising their rights and for this reason require special attention.[123] As a result, specific international human rights treaties have been developed starting with the Convention on the Elimination of All Forms of Racial Discrimination (1965) and it has been followed by other instruments dealing with the specific rights of and obstacles for certain groups such as women (1979), migrant workers (1990) and the disabled (2006). The adoption of these specialised human rights conventions testifies to the fact that states are aware of the special attention that these groups need due to their

120 Ibid. p. 13 – 24.

121 Lourdes Peroni and Alexandra Timmer, 'Vulnerable Groups: The Promise of an Emerging Concept in European Human Rights Convention Law' (2013) 11 International Journal of Constitutional Law 1056. p. 1061.

122 Alexander HE Morawa, 'Vulnerability as a Human Rights Law' (2003) 6 139. p. 147.

123 Such as the Committee on the Rights of the Child, the Committee on the Elimination of Discrimination against Women and the Committee on the Elimination of Racial Discrimination.

vulnerability.[124] But also within general human rights instruments, references have been made to the particular situation of specific groups.[125]

Different regional instruments have been introduced since the 1950s. And these have also embodied the idea of vulnerable groups.[126] The European Convention on Human Rights is one these instruments. Connected to this instrument the European Court of Human Rights has recognised particular vulnerable groups in the European context: the Roma population, mentally disabled persons, asylum seekers, and people living with HIV.[127] In the Inter-American system the concept of vulnerability has a central role, both explicitly and implicitly, in the reasoning of the Court as well as the Commission.[128] 'The Court systematically deduces a special need for protection tailored according to the vulnerabilities of certain groups which are particularly targeted. Children, indigenous peoples, women, and migrants are all such groups through whose lens the Court will interpret the extent and nature of the Convention's rights.'[129] An example is the *Rosendo Cantu v. Mexico case* where it was found that Rosendo Cantu, as a young girl, lived in a situation of vulnerability as she was a minor and was part of an indigenous community.[130] The regional human rights courts assist in further developing the concept of vulnerability within the human rights framework by identifying vulnerable groups; however, a clear definition is also lacking as far as these institutions are concerned.

Vulnerability is not the only term which is used to signify groups and individuals in need of special protection. Within the international human rights framework different terms are used to signify individuals and groups which need targeted measures to support them in exercising their rights, such as those who are marginalised, stigmatised, disadvantaged and discriminated.[131] This study uses the

---

124 Ibid. p. 139.

125 See for instance Article 7 (1) ICESCR paragraph 1: 'Fair wages and equal remuneration for work of equal value without distinction of any kind, in particular women being guaranteed conditions of work not inferior to those enjoyed by men, with equal pay for equal work.'

126 Alexandra Timmer, "A Quiet Revolution: Vulnerability in the European Court of Human Rights", see Peroni and Timmer.

127 The concept of vulnerability within the system of the European Court of Human Rights was introduced in Chapman v. the United Kingdom. Here vulnerability was articulated in relation to the Roma minority: 'belonging to a group (in this case, the Roma minority) whose vulnerability is partly constructed by broader societal, political, and institutional circumstances (in this case, power differentials and a planning framework unresponsive to the needs arising from a way of life different from that of the majority.)' (Peroni & Timmer, 2013) p. 1063.

128 Ludovic Hennebel, 'The Inter-American Court of Human Rights: The Ambassador of Universalism' (2011) Special ed Quebec Journal of International Law 57. p. 62.

129 Ibid. p. 64.

130 'In general, the indigenous populations live in a situation of vulnerability and this is reflected in a number of ways, such as in the administration of justice and health care services, and in particular because they do not speak Spanish and have no interpreters, because they lack the financial resources to find a lawyer, to travel to health care centres or to the courts and also because they are often victims of abusive practices or practices that violate due process.' Paragraph 70.

131 For instance in the 2008 Guidelines of the ESCR Committee the terms disadvantaged and marginalized are used. See for instance Article 3 CESCR, 'Guidelines on Treaty-Specific Documents

term and the subject of vulnerability for several reasons which have to do with the position of this term in the field of water management and within human rights. Within the human rights approach the protection of the vulnerable is one of the central commitments and is highly relevant in relation to the human right to water.[132] Firstly, because those who currently lack access to water are predominantly vulnerable groups.[133] Due to the Millennium Development Goals the global community at large, and states individually, have made efforts to increase access and aim to attain universal access to water. Different targets have been reached.[134] However, it seems that those targets were what can be described as 'low hanging fruit'[135], meaning that those who were easiest to reach were targeted in order to achieve the required percentage. As a result, the most disadvantaged and vulnerable groups have missed out on these developments.[136] Consequently, when discussing improving access to water and realising the right to water universally, it is important to recognise that those who currently lack access are predominantly vulnerable and disadvantaged groups who are not easily reached and are, as a result, mostly dependent on traditional sources of water.

States have the immediate obligation to adopt measures to combat inequalities in access and prioritising access for vulnerable individuals should be a focal point in the policies and practices of states.[137] Secondly, the situation of vulnerable groups serves as a litmus test for the efficacy of the human right to water in a state.[138] And, thirdly, by 2030 the world is expected to have a 40% shortfall in water supply if the current trend continues.[139] This means that the number of people who lack access to water may increase. If current water management will be able to adapt in order to provide access to water for currently vulnerable groups, it can be assumed that water management will

---

to Be Submitted by States Paries under Articles 16 and 17 of the International Covenant on Economic, Social and Cultural Rights' 15.

132 Chapman and Carbonetti. p. 683.

133 'It is therefore people living in poverty, and particularly marginalised and vulnerable groups and individuals, who are most likely to lack access to safe water and sanitation. These groups are also less likely to enjoy access to adequate housing, health and education, in part due to this lack of access to water and sanitation.' Catarina De Albuquerque and Virginia Roaf, *On The Right Track. Good Practices in Realising the Rights to Water and Sanitation* (2012th edn, 2012) <www.ohchr.org/Documents/Issues/Water/BookonGoodPractices_en.pdf>. p. 21.

134 R Quentin Grafton and others, *Global Water: Issues and Insights* (ANU Press 2014). p. 85.

135 Stephen C Mccaffrey, 'Human Right to Water: A False Promise' (2016) 47 University of the Pacific Law Review 221. p. 6.

136 Albuquerque. p. 36.

137 De Albuquerque and Roaf, 'Principles'. p. 15.
The obligation not to discriminate is an immediate obligation. Chapter 5 will explore the question of prioritization.

138 Joshua Castellino, 'The MDGs and International Human Rights Law: A View from the Perspective of Minorities and Vulnerable Groups' (2009) 13 The International Journal of Human Rights 10 <www.tandfonline.com/doi/abs/10.1080/13642980802532275> accessed 16 December 2013. p. 23.

139 UNEP, *Option for Decoupling Economic Growth from Water Use and Water Pollution* (UNESCO 2015). p. 17.

also be able to deal with the expected future increase in vulnerable individuals due to climate change and other developments.[140]

The use of the concept of vulnerability should not overlook links between the different terms within the human rights framework. Especially the connection between vulnerable groups and the principle of non-discrimination is highly important. Discrimination, historically or deeply engrained, can be the main reason for the vulnerability of individuals and groups.[141] The CESCR provides in General Comment No. 20 on non-discrimination in economic, social and cultural rights a list which could be seen as the most comprehensive list of those who can be considered to be vulnerable. Among them are refugees, women and migrant workers. This Comment identifies prohibited grounds of discrimination such as religion and national origin.[142] It explains, however, that this list is not exhaustive[143] and also recognises that individuals and groups may be discriminated against on multiple grounds at the same time[144], also referred to as intersectionality.[145] The Committee also points to the specialised human rights instruments which deal with the challenges experienced by specific groups such as the Convention on the Rights of Persons with Disabilities. For this reason, measures such as affirmative action, participation and the redistribution of power and resources are vital.[146] What can be concluded is that vulnerability does not have a clear definition[147] but it could be said that human rights committees and courts consider individuals and groups to be vulnerable when, on grounds beyond their control, they are unable to realise rights for themselves with the means they have available.[148]

The use of the term vulnerability has not been uncontested. This categorisation has received criticism as the categories by themselves can marginalise and stigmatise communities and individuals belonging to these groups. These groups can be seen as

---

140 See Laura Westra, 'Climate Change and the Human Right to Water' (2010) 1 Journal of Human Rights and the Environment 161 <http://e-elgar.metapress.com/openurl.asp?genre=article&id=doi:10.4337/jhre.2010.02.02>.

141 UNEP, 'Vulnerability of People and the Environment – Challenges and Opportunities' (J Jager and Kok eds, 2007) <www.jci.org/cgi/content/abstract/115/10/2656%5Cnwww.tandfonline.com/doi/full/10.3109/09553002.2015.1047989%5Cnwww.pubmedcentral.nih.gov/articlerender.fcgi?artid=3820435&tool=pmcentrez&rendertype=abstract%5Cnwww.nature.com/doifinder/1>. p. 69.

142 General Comment No. 20 paragraph 16 – 35.

143 General Comment No.20 par. 20.

144 General Comment No.20 par. 17.

145 See Aisha Nicole Davis, 'Intersectionality and International Law: Recognizing Complex Identities on the Global Stage' (2015) 28 Harvards Human Rights Journal 205. and Johanna E Bond, 'International Intersectionality: A Theoretical and Pragmatic Exploration of Women's International Human Rights Violations' (2003) 52 Emory Law Journal 71.

146 De Albuquerque and Roaf, 'Principles'. p. 13.

147 Another UN instrument, The Habitat Agenda, qualifies vulnerable groups and individuals as those who are marginalized and excluded 'from the socio-economic mainstream and decision-making processes and the lack of access on an equal basis to resources and opportunities', section IV B, 4, para. 93.

148 Ssenyonjo. p. 36.

'lesser than' and 'deviant'.[149] Several examples can be found where the designation of vulnerability has resulted in problematic responses by the state. This designation can for instance result in strict surveillance and patronising behaviour by the state.[150] It can also be a way for states to disguise their unwillingness to change their practices.[151] Another point of criticism is that the designation clusters groups together without looking at the different causes of vulnerability and whether these groups are in fact truly vulnerable. Also, labelling certain groups as vulnerable creates the impression that other groups are not vulnerable. Still, vulnerability is intrinsically connected to the human experience. The emphasis, however, should be on the situation which makes individuals and communities vulnerable, rather than assigning the term to entire groups and assuming that they need specific protection.[152]

2.3 VULNERABILITY IN RELATION TO THE HUMAN RIGHT TO WATER

The human right to water has been authoritatively interpreted in General Comment No. 15 of the Committee on Economic, Social and Cultural Rights. Therefore the Committee's views are central to this research.[153] This section will discuss the views of the Committee on the characteristics of vulnerable groups and seeks to expand on these views for the purpose of the research. So far, however, the Committee has not provided a specific definition of 'vulnerable groups' or 'vulnerable individuals'. The Committee does not consistently use the term 'vulnerable'. It apparently uses this term interchangeably with terms such as disadvantaged and marginalised.[154] It also does not provide a list of which groups can be considered vulnerable or the characteristics of vulnerability which the Committee agrees upon.[155] Nevertheless, there are a few indications about vulnerability that can be derived from the Committee's work.

Firstly, it seems that the Committee characterises vulnerable people as those who are at risk of having their rights infringed or having difficulty in exercising their rights. For instance, in General Comment No. 15 on the human right to water, it states that specific groups 'have traditionally faced difficulties in exercising this right.'[156] Secondly,

---

149 Martha Albertson Fineman, 'Equality, Autonomy, and the Vulnerable Subject in Law and Politics' in Martha Albertson Fineman and Anna Grear (eds), *Vulnerability: reflections on a new ethical foundation for law and politics* (2nd edn, Routledge 2016). p. 16.

150 Ibid. p. 16.

151 Morawa. p. 148.

152 (De Albuquerque & Roaf, 2014a) p. 18.

153 The protection of the rights and interests of vulnerable groups has been a recurring theme in the work of the ESCR Committee Chapman and Carbonetti. p. 685.

154 Ibid. p. 689.

155 (Chapman & Carbonetti, 2011) p. 683.

156 Committee on Economic Social and Cultural Rights. par. 16.

the Committee does recognise that certain groups and individuals[157] can be deemed to be vulnerable. It recognises this in relation to specific rights and situations. In a few General Comments specific groups have been identified as being at risk in relation to a particular right.[158] For instance, in General Comment No. 12 on the right to adequate food, the Committee identifies socially vulnerable groups such as landless persons and the poor.[159] It also mentions physically vulnerable individuals like infants and the physically disabled.[160] In relation to the human right to water the Committee asks for special attention to be given to certain groups in General Comment No. 15. Among these groups are women, children, refugees and migrant workers. These groups can be seen as vulnerable and the Committee encourages states to take specific measures to target them.

Besides the authoritative views of the CESCR, the former Special Rapporteur[161] has also weighed in on the topic and explains that:

> *Vulnerability often refers to individuals or groups under threat of physical or mental harm, for example, at times of conflict, abuse, rape or neglect, and perhaps because of their disadvantaged social or economic status. All individuals may sometimes be vulnerable, requiring help to realise their human rights. By emphasising the situation that makes individuals vulnerable, rather than assigning entire populations groups to that status, the concept of vulnerability becomes less patronising and victimising.*[162]
>
> *Vulnerable individuals and groups include people who have specific needs with respect to water and sanitation, and who are not able to meet those needs themselves. This may include pregnant women, new mothers, children, older persons, persons with disabilities, and those living through difficult circumstances due to disasters or climate change. Vulnerability is not necessarily life-long – people can move in and out of being 'vulnerable'.*[163]

What can be observed is that, unlike the CESCR, the Special Rapporteur does take environmental factors into account, even though only to a limited extent. Environmental factors including spatial and geographic attributes are other causes of vulnerability which render certain groups and individuals particularly vulnerable in relation to accessing water for domestic purposes.[164] The environment in which

157 There is ambiguity concerning whether the jurisprudence of the Committee applies to individuals, groups or both (Chapman and Carbonetti. p. 690.

158 Ibid. p. 724.

159 General Comment No. 12 par. 13.

160 General Comment No. 12 par. 13.

161 The Special Rapporteur on the human rights to safe drinking water and sanitation reports to the Human Rights Council.

162 De Albuquerque and Roaf, 'Principles'. p. 18.

163 Catarina De Albuquerque and Virginia Roaf, 'Sources', *Realising the human rights to water and sanitation: A handbook by the UN Special Rapporteur Catarina De Albuquerque* (2014). p. 10.

164 Leila M Harris, 'Water Rich, Resource Poor: Intersections of Gender, Poverty, and Vulnerability in Newly Irrigated Areas of Southeastern Turkey' (2008) 36 World Development 1 <http://linkinghub.elsevier.com/retrieve/pii/S0305750X08001319> accessed 18 November 2013. p. 3 'e.g. upstream or downstream users, or in relation to the particular hydro-geography.' See also Farhana Sultana, 'Water, Water Everywhere, But Not a Drop to Drink: Pani Politics (Water Politics) in Rural Bangladesh'

individuals and groups live is closely related to how and whether their human right to water can be realised, especially those who depend on local and traditional water resources as most of the currently unserved do.[165]

For instance, if local water resources are scarce and communities experience drought, water for domestic purposes needs to be transported from other locations. Another example of the connection with environmental factors can be seen when local communities have their water resources contaminated by chemicals or other harmful substances. What should be taken into account in analysing this connection is whether individuals and groups have other options for obtaining access to safe water if traditional sources are unsafe. Do they, for instance, have enough financial means to buy water if a river becomes contaminated or is there sufficient infrastructure available which facilitates access? If not, fulfilling the human right to water of these groups and individuals who depend on traditional water resources remains weak.

Even though the Committee does not directly link environmental factors to vulnerability, it does discuss certain geographical and environmental aspects related to the human right to water. In General Comment No. 15 it mentions that rural and deprived areas should have access to proper water facilities. These areas for instance include informal human settlements. It also demands that states protect water resources from contamination.[166] However, it does not recognise environmental factors as characteristics of vulnerability. In this regard, terms such as 'environmental vulnerability' can provide a useful analysis. Environmental vulnerability is used to describe the degree to which individuals and groups are susceptible to environmental damage.[167]

What should be kept in mind is that there are different definitions of (environmental) vulnerability used by different disciplines. And that within these disciplines not only persons, but also systems[168], states and societies can be considered vulnerable. During the past decade this concept has mostly been used in climate change research and in relation to natural disasters.[169]

The UNEP defines vulnerability as follows:

'Vulnerability is a function of sensitivity to present climatic variability, the risk of adverse future climate change and capacity to adapt. The extent to which climate change may damage

---

(2007) 9 International Feminist Journal of Politics 494 <www.tandfonline.com/doi/abs/10.1080/14616740701607994>.

165 For instance, in relation to indigenous groups who depend on traditional sources of water. 'Depletion and pollution of water sources for development by non-native enterprises and individuals have often limited the ability of native peoples to carry on water-dependent vocations like farming and fishing, as well as to perpetuate cultures that may depend on traditional subsistence activities and spiritual practices requiring water.' David H Getches, 'Indigenous Peoples' Rights to Water Under International Norms' (2005) 16 Colorado Journal of International Environmental Law and Policy 259. p. 260.

166 Committee on Economic Social and Cultural Rights. par. 8.

167 Barnett. p. 257.

168 W Neil Adger, 'Vulnerability' (2006) 16 Global Environmental Change 268. p. 273.

169 See for instance Barnett. on human rights and vulnerability to climate change.

or harm a system; vulnerability is a function of not only the systems' sensitivity, but also its ability to adapt to new climatic conditions.'

The environmental vulnerability of individuals and groups is closely linked to the vulnerability of ecosystems.

'Environmental vulnerability is related to pollution and water shortages aggravated by climate variability where rapid-onset extreme hydrometeorological events (flash floods, cyclones) and medium-term (drought, erosion) and long-term (desertification) processes aggravate fragile soil conditions in areas of steep slopes and stony and shallow layers with a superficial horizon of fertility.'[170]

The concept of vulnerability has become an important part of socio-ecological system analysis and sustainable livelihood research as disaster and entitlement theories[171] have contributed to the development of the concept.[172] Because of the contributions that these disciplines have made the concept has become dynamic. Nevertheless, social factors as contributing and often leading factors have largely not been dealt with in certain disciplines.[173] The human rights framework could also include this perspective so as to further develop an integrative approach to the concept.[174]

Including environmental factors in identifying vulnerable groups and individuals will assist in recognising that the (quality of the) environment in itself can be another cause of vulnerability and it can exacerbate other forms of vulnerability. When environmental vulnerabilities increase, these have a negative outcome for people and, in turn, influence their social vulnerability and the degree to which they enjoy human rights.[175] The same can be said vice versa: the degree to which individuals enjoy human rights is strongly connected to their capacity to adapt when environmental damage occurs.[176] The

170 Úrsula Oswald Spring, 'Social and Environmental Vulnerability in a River Basin of Mexico' in Úrsula Oswald Spring, Hans Günter Brauch and Keith G Tidball (eds), *Expanding Peace Ecology: Peace, Security, Sustainability, Equity and Gender*, vol 12 (Springer International Publishing 2014) <http://link.springer.com/10.1007/978-3-319-00729-8> accessed 22 December 2013. p. 87.

171 One of the challenges of using terms across disciplines is whether the term has the same meaning in various disciplines. For instance, in vulnerability in relation to food insecurity entitlement theory is used. While entitlement in legal theory and therefore human rights refers to rights it can also mean: 'actual or potential resources available to individuals based on their own production, assets or reciprocal arrangements.' Adger. p. 270.

172 Ibid. p. 269.

173 Karina Vink, Kuniyoshi Takeuchi and Kelly M Kibler, 'A Quantitative Estimate of Vulnerable People and Evaluation of Flood Evacuation Policy' (2014) 9 Journal of Disaster Research 887. p. 888.

174 'Integrative frameworks view vulnerability as depending on both biophysical and human factors.' Jesse Ribot, 'Vulnerability Does Not Fall from the Sky: Toward Multiscale, Pro-Poor Climate Policy' in Andrew Norton and Robin Mearns (eds), *Social Dimensions of Climate Change: Equity and Vulnerability in a Warming World* (The World Bank 2010). p. 53.

175 (Oswald Spring, 2014) p. 93.

176 In relation to this the concept of resilience can be useful as it explains how a system can absorb shocks like natural disasters but it can also be applied to environmental pollution by human causes. See Christopher Hobson, 'Human Security after the Shock: Vulnerability and Empowerment' in Christopher Hobson, Paul Bacon and Robin Cameron (eds), *Human Security and Natural Disasters*

focus of this study is the combination of these categories of vulnerability together, i.e., environmental and social vulnerability. They determine the obstacles for individuals and groups which are unserved and underserved in realising their human right to water and which measures can be taken for these groups so that they can realise this right.

There are several benefits of including environmental factors when assessing the realisation of the human right to water. The benefit of including environmental factors in the human rights framework is that those groups who are disproportionately affected by environmental degradation fall within the scope of targeted measures. Also, the challenge of vulnerability is that it is not necessarily concerned with the physical scarcity of resources but with access to these resources.[177] Therefore environmental factors should be considered in combination with social vulnerability as defined by the UN Committee on ESCR. Because even in situations where water is abundant, certain groups and individuals are still vulnerable to having their access to water restricted regardless of the availability of water in general.

Vulnerable groups should be identified in the national context taking the groups suggested by the CESCR into account.[178] The categories provided by the Committee do not apply in and of themselves. In particular national contexts, particular groups can be considered to be vulnerable groups especially considering the combination of social and environmental factors.

2.4 DEFINING VULNERABLE GROUPS FOR THIS STUDY

Taking what has been previously discussed into account, this study looks at specific groups of vulnerable individuals. The term vulnerable has been chosen, as opposed to disadvantaged or marginalised, due to its cross-disciplinary use. Similar to the issue of access to water, the term vulnerability appears to be relevant within the intersection between environmental science and human rights.[179] The term vulnerable groups in this context refers to those groups that are susceptible to not being able to obtain sustainable access to adequate quantities of water of an acceptable quality for domestic purposes due to a combination of social and environmental factors.[180] As discussed, this

(Routledge/ Taylor & Francis 2014) <http://isp.unu.edu/research/human-security/index.html>. & Mansour Haidera and others, 'Water Scarcity and Climate Change Adaptation for Yemen's Vulnerable Communities' (2011) 16 Local Environment 473 <www.tandfonline.com/doi/abs/10.1080/13549839.2011.565465> accessed 6 November 2013. & Barnett. p. 259.

177 Leon Hermans, 'Climate Change, Water Stress, Conflict and Migration. Taking Stock of Current Insights through a Vulnerability Lens', *Climate change, water stress, conflict and migration* (2012). p. 23.

178 For instance the Brasilia Regulations recognize that: 'The specific definition of vulnerable people in each country will depend on their specific characteristics, and even on their level of social or economic development.' Section 2 paragraph 4.

179 See for instance Barry Smit and Johanna Wandel, 'Adaptation, Adaptive Capacity and Vulnerability' (2006) 16 Global Environmental Change 282 <http://linkinghub.elsevier.com/retrieve/pii/S0959378006000410> accessed 19 March 2014.

180 Developed from the term 'water security' (UN-Water, 2013). Water security is defined as "the capacity of a population to safeguard sustainable access to adequate quantities of acceptable quality water for

definition expands on the concept of vulnerability adopted by the CESCR. It analyses the situation of those who do not have access to water because of social factors and who are, at the same time, unable to depend on their environment to meet their needs.

Some may argue that the consideration of environmental factors within the human rights framework goes beyond the scope of this framework. However, certain human rights are closely connected to the environment and the fact that environmental protection has now become a part of human rights considerations has become widely accepted.[181] It can even be argued that international environmental law has been broadening the substantive scope of economic, social and cultural rights and vice versa. Adding environmental factors to the concept of vulnerability as discussed within the human rights framework follows this trend of expansion. Several examples of this cross-over of environmental concerns to human rights can be found throughout the past few decades.[182] Even though a human right concerning the environment has not been formulated in a legally binding international instrument, the constitutions of around 100 states do recognise such a right, both indirectly and directly.[183] Also soft law international instruments discuss environmental rights. For instance, the 1994 Draft Principles on Human Rights and the Environment and the Declaration on Human Rights and Climate Change, but also the human right to water itself, as will be discussed in chapter 3, are clear examples of international environmental law influencing international human rights law. In addition, the CESCR makes use of international environmental law when formulating its (non-legally binding) General Comments.[184]

Closely connected to environmental concerns is climate change. Climate change has several characteristics, one of them being that water stress will worsen in many parts of the world due to changes in rainfall patterns, glacial and snow melt and rising salinity in low-lying coastal areas.[185] There is an increased understanding about the relationship between climate change and the human rights framework.[186] Climate change acts as an multiplier of existing vulnerabilities.[187] However, the categories of vulnerable groups within the climate change framework, even though there is an overlap, also differ from

sustaining livelihoods, human well-being, and socio-economic development, for ensuring protection against water-borne pollution and water-related disasters, and for preserving ecosystems in a climate of peace and political stability."

181 See for instance Avadne Grant, 'International Court and Environmental Human Rights: Re-Imagining Adjudicative Paradigms' in Anna Grear and Louis J Kotze (eds), *Research Handbook and the Environment* (Edward Elgar Publishing 2015). on the 'greening' of human rights by international courts.

182 The 1994 UN Draft Principles on Human Rights and the Environment are a clear example of the nexus between human rights and environmental rights.

183 See James R May and Daly, 'Emerging Environmental Constitutionalism', *Global Environmental Constitutionalism* (Cambridge University Press 2015); David R Boyd, 'Constitutions, Human Rights and the Environment: The Context', *The Environmental Rights Revolution* (UBC Press 2012).

184 See for example General Comment no. 12: The Right to Adequate Food.

185 Robin Mearns and Andrew Norton, 'Equity and Vulnerability in a Warming World: Introduction and Overview' in Robin Mearns and Andrew Norton (eds), *The Social Dimensions of Climate Change Equity and Vulnerability in A Warming World* (The World Bank 2010). p. 15.

186 See Richard P. Hiskes, *The Human Right to a Green Future* (Cambridge University Press 2009).

187 Mearns and Norton. p. 2.

those highlighted in the human rights framework. The poor, women and indigenous peoples are vulnerable groups in both frameworks. Nevertheless, the climate change framework recognises additional groups which clearly embody geographical characteristics such as people in rural drylands, the urban poor and children.[188]

A clear benefit of this expansion of the scope of economic, social and cultural rights is that it also opens up the use of other enforcement mechanisms outside the human rights framework for the protection of human rights.[189] This gives vulnerable groups more options for having their human rights in general and the right to water specifically protected and realised as will be discussed in section 2.5.

2.5 OBLIGATIONS AND CONSEQUENCES RELATED TO THE DESIGNATION OF VULNERABILITY

The classification of vulnerability brings along consequences and responsibilities for the state. From the General Comments of the ESCR three types of obligations in relation to vulnerable groups can be discerned. Firstly, states need to monitor the status of these individuals and groups. Secondly, the basic economic and social rights of vulnerable groups need to be protected and, lastly, they should be given priority in the realisation of human rights.[190]

In order to implement targeted measures, vulnerable groups need to be identified and the measures and their effects need to be monitored. For this reason, data should be aggregated according to specific group traits such as gender, age, ethnicity etc. States are obligated to provide disaggregated data on the position of minorities. This is often not done[191] and as a result the inequalities experienced by these groups remain invisible.[192] Also, access to water and sanitation are often measured in a manner which does not show the actual situation.[193] What should be kept in mind however is that data collection brings up questions around privacy and stigmatisation.[194] Identifying which groups and individual can be considered vulnerable can be seen as inappropriate. However, data collecting provides insight into the social aspects which influence vulnerability. As will be argued, this information should be considered in conjunction with other information as by itself data about social aspects is not conclusive in determining who has access to water.

---

188 Ibid. p 21.

189 See Chapter 4.

190 Chapman and Carbonetti. p. 726.

191 One of the reasons is that there is only one monitoring system which requires direct reporting Castellino. p. 23.

192 Ibid. p. 16.

193 Karina Vink, 'Transboundary Water Law and Vulnerable People: Legal Interpretations of the "equitable Use" Principle' [2014] Water International 1 <www.tandfonline.com/doi/abs/10.1080/02508060.2014.951827> accessed 16 September 2014. p. 6.

194 Office of the High Commissioner, *Human Rights Indicators A Guide to Measurement* (2012). p. 22.

Discrimination can be a major obstacle for realising the human rights of vulnerable people. The fundamental principle of non-discrimination and equality is considered to be an 'immediate and cross-cutting obligation.'[195] Article 2(2) ICESCR further explains that the rights mentioned in the Convention must be exercised without any kind of discrimination. This means that states must end *de jure* and *de facto* discrimination which is caused by government actions. In addition, past systemic discrimination must be addressed.[196] This may require the adoption of preferential measures and affirmative action targeting specific groups in certain circumstances.[197] For vulnerable groups this means that specific attention should be paid to their situation.[198]

The human rights approach requires that systemic and underlying causes of discrimination are understood and addressed. For this reason, tackling the issues of water access goes beyond merely looking at who does not have access to water. It deals with social power relations which are reflected in law, policies and institutional structures as well as social and cultural practices.[199] These underlying structural causes affect access to water and targeted measures following their analysis can effectively deal with them. This enhances sustainable solutions for increasing access to water and working towards universal access.

Considering environmental factors as a cause of vulnerability shows that discrimination is not always or by itself the root cause of individuals or groups not having access to water. Therefore merely discussing the non-discrimination principle would be too limited a perspective in relation to the human right to water even though in some instances discrimination and the pollution of the environment are strongly connected.

The connection between discrimination and environmental degradation can be explained by the term 'environmental justice' which refers to the movement to redress a disproportionately adverse environmental impact on vulnerable groups[200] and has its origin in human rights.[201] It focusses on inequalities based on ethnic and/or social characteristics in the distribution of environmental benefits and harm.[202] For instance, this framework explains that vulnerable groups are most often exposed to environmental disasters and pollution as '*they tend to live where land is cheap or free because it is exposed to climatic extremes such as flood, drought and land- slides, or is unproductive (...). For example, pastoral populations are highly vulnerable to hunger in*

195 General Comment No. 20 par. 7.
196 Winkler, *The Human Right to Water*. p. 222.
197 Chapman and Carbonetti. p. 694.
198 Emilie Filmer-Wilson, 'The Human Rights-Based Approach to Development: The Right to Water' (2005) 23 Netherlands Quarterly of Human Rights 213. p. 236.
199 (I.T. Winkler, 2015) p. 222.
200 Adjusted from Robert William Collin and Robin Morris Collin, 'Sustainable Development Environmental Justice and Sustainability' in Michael Redclift (ed), *Routledge International Handbook of Sustainable Development* (Taylor and Francis 2015). p. 209.
201 Richard Filcak, 'On the Distribution of the Environment Benefits and Adverse Impacts: The Environmental Justice and Vulnerability of People in Risk' (2009) 1 Environmental Law Review of Eastern & Central Europe 63 <www.prog.sav.sk/subory/esee/8_vulnerability_paper.pdf>.p. 3.
202 Ibid. p. 1.

times of drought when there are restrictions placed on their freedom of movement.'[203] And poor communities tend to live along riverbanks and on floodplains.[204]

Vulnerability is experienced locally. However, one challenge which the consideration of environmental factors brings is that the solution might not be found at the state level alone. The causes of vulnerability may occur at different social, geographical and temporal levels.[205] Pollution, for instance, may be a transboundary issue.[206] Identifying the causes of vulnerability helps to identify at which level, and by implementing which measures, interventions can be effectively made. Policies which aim to reduce or compensate vulnerability are implemented through state institutions.[207] And these institutions bear responsibility for meeting the needs of multiple subgroups whose interests and needs may compete when laying claim to resources.[208]

Merely looking at vulnerability caused by discrimination would omit certain individuals and groups who could be helped by targeted measures (both directly as well as indirectly). Prohibited grounds of discrimination provide for cases of alleged discrimination to be (easily) adjudicated but make this challenging for cases which fall outside the scope of these grounds.[209] For instance, if in one area two communities are residing, one large indigenous community and one small non-indigenous community, and in this area adequate infrastructure is lacking because the government prefers not to invest in a minority-dominated community, then only the indigenous community would be able to rely on the principle of non-discrimination. As a result, access to water services in general could be improved, thereby also reaching the non-indigenous community. Therefore, indirectly, the non-indigenous community would have their human right to water realised. This is nevertheless not the ideal situation as affected communities should have access to justice themselves. This example illustrates why it is necessary for the concept of vulnerability not to simply mirror the prohibited groups under the non-discrimination principles but to widen its application.[210]

A way in which the rights and entitlements of vulnerable groups can be identified and taken up is by ensuring their participation.[211] Individuals and groups who are affected by decisions concerning water allocation must be able to participate in the decision-making process; this includes vulnerable individuals and groups.[212] The assumption is that the requirements of non-discrimination and participation (and

---

203 (Barnett, 2010) p. 259.

204 Mearns and Norton. p. 20.

205 Ribot. p. 64.

206 We could also think of water stress as a result of climate change. Barnett. p. 266.

207 Ribot. p. 64.

208 Ibid. p. 64.

209 Romina I Sijniensky, 'From the Non-Discrimination Clause to the Concept of Vulnerability in International Human Rights Law' in Yves Haeck and others (eds), *The Realisation of Human Rights: When Theory Meets Practice* (Intersentia 2014). p. 268.

210 Ibid. p. 263.

211 Participation will be discussed more in depth in Chapters 4 & 5.

212 Kiefer and others, *The Human Right to Safe Drinking Water and Sanitation in Law and Policy – a Sourcebook*. p. 12.

accountability) will have the combined effect of empowering vulnerable groups.[213] Besides participation during water policy development, guaranteeing access to justice for vulnerable groups plays an important role in realising the right to water. The Brasilia Regulations Regarding Access to Justice for Vulnerable Groups require states to take many factors into account, such as gender and social and economic factors; however, again they do not take into account environmental factors.[214]

2.6 PROTECTION OF WATER ACCESS FOR PARTICULAR VULNERABLE GROUPS

This chapter argues for a wider scope to be given to the concept of vulnerability in order to effectively protect access to water for vulnerable groups. Vulnerable groups and individuals are considered to be those groups and individuals who are susceptible to not being able to obtain sustainable access to adequate quantities of water of an acceptable quality for domestic purposes due to the combination of social and environmental factors. The previous sections have explained why the definition needed to be expanded and that environmental aspects should be included. Nevertheless, in the current framework, only social vulnerability is taken into account. What should be kept in mind is that these categorisations can be helpful in the protection of certain groups, but the challenge is that this approach often encourages dealing with problems in a vacuum. In addition, the treaties make it seem that the identities of vulnerable groups are one dimensional instead of recognising that these groups often have complex identities.[215]

Access to water affects different groups differently. Vulnerable groups are disproportionately affected by inequalities in water access. Because of this, certain human rights instruments which target specific groups include provisions which aim to protect access for groups, children and disabled people who may have to depend on the care of others and therefore experience difficulties in accessing water independently.[216] The following section discusses the explicit recognition of several vulnerable groups' right to water. This is followed by a discussion of other instruments and provisions which allow for rights in relation to water to be applicable to vulnerable groups.

Access to water is crucial for everyone. Nonetheless, there are several vulnerable groups for whom the right to water is specifically recognised. Even though the ICESCR

---

213 Catarina De Albuquerque, 'Report of the Independent Expert on the Issue of Human Rights Obligations Related to Access to Safe Drinking Water and Sanitation "Compilation of Good Practices"' (2010). paragraph 37.

214 This instrument is not legally binding. However, it is considered to be one of the most successful attempts at providing a comprehensive definition of vulnerable groups Francesca Ippolito and Sara Iglesias Sánchez, 'Introduction' in Francesca Ippolito and Sara Iglesias Sánchez (eds), *Protecting Vulnerable Groups. The European Human Rights Framework* (Hart Publishing 2015). p. 1.

215 See Davis, 'Intersectionality and International Law: Recognizing Complex Identities on the Global Stage'.

216 World Health Organization, 'GLAAS 2012 Report. UN-Water Global Analysis and Assessment of Sanitatioin and Drinking-Water'. p. 44 Sijniensky. p. 268.

and the ICCPR lack explicit rights to water, there are other human rights instruments with limited scope where the right to water is explicitly mentioned.

One of the reasons for explicitly recognising specific groups' right to water is that there are specific situations which warrant a different amount of water availability for certain groups. Not all groups require the same amount of water. This can depend on several factors such as health, age and gender. Supplying everyone with the same amount of water may not meet everyone's needs.[217] Paragraph 12 of General Comment No. 15 recognises this as well: 'Some individuals and groups may also require additional water due to health, climate, and work conditions.'[218]

The following subsections discuss the rights to water for women, the disabled and children. All of the three conventions discussed have received a large number of ratifications, headed by the CRC with 193 states parties[219], and should therefore be implemented almost universally. This section also provides an interpretation of the right to water for indigenous peoples as for them rights to land and water resources are often protected. Lastly, this section will discuss the vulnerability of future generations as they are a specific group in relation to sustainability which could require protection.

2.6.1 1979 CONVENTION ON THE ELIMINATION OF ALL FORMS OF DISCRIMINATION AGAINST WOMEN

A lack of access to water affects women and girls disproportionately. Women and girls often bear the burden of fetching water which can take up a significant part of the day. Rural women, for instance, spend almost four to five hours a day doing this. This lost time could have been spent undergoing education or undertaking other activities. Also, as they collect water from unprotected sources women run health risks as they are exposed to unsafe water. With regard to the lack of sanitation, women and girls face difficulties as they have different physical needs than men. They are at risk of experiencing sexual violence when they have to walk long distances to have enough privacy for using sanitation facilities. Besides women's dependence on water resources, women also play a vital role in the management and safeguarding of these resources. They both use and provide water, yet their contribution is often not recognised in institutional arrangements.[220]

As explained, women in general are disproportionately affected by a lack of access to water. However, some women are more vulnerable to having their right to water infringed. Due to these reasons, CEDAW places water supply in the context of rural development and applies it specifically to rural women. Article 14(2)(h) of the 1979

217 Vink. p. 2.
218 (Committee on Economic, Social and Cultural Rights, 2002) paragraph 12.
219 Records of treaty ratifications including the CRC are available at: http://treaties.un.org/Pages/Treaties.aspx?id=4&subid=A&lang=en last visited in December,2012.
220 Principle 3 of the Dublin Statement.

Convention on the Elimination of All Forms of Discrimination against Women (CEDAW) links the right of water supply to adequate living conditions. It states:

1. *States Parties shall take into account the particular problems faced by rural women and the significant roles which rural women play in the economic survival of their families, including their work in the non-monetized sectors of the economy, and shall take all appropriate measures to ensure the application of the provisions of the present Convention to women in rural areas.*
2. *States Parties shall take all appropriate measures to eliminate discrimination against women in rural areas in order to ensure, on a basis of equality of men and women, that they participate in and benefit from rural development and, in particular, shall ensure to such women the right:*
   (h) To enjoy adequate living conditions, particularly in relation to housing, sanitation, electricity and water supply, transport and communications.

This CEDAW provision is restricted in its application. It only protects a very specific group of women, those living in rural areas. For these women access to water is critical for livelihood activities and can increase a household's income and food security.[221] This provision implicitly recognises geographical and environmental factors which negatively affect access to water for these women. General Recommendation No. 34 on the rights of rural women explains that 'environmental issues, including climate change and natural disasters, often provoked by unsustainable use of natural resources, as well as poor waste management practices, also have detrimental impacts on the wellbeing of rural women.'[222]

As a result, 'states parties should address specific threats posed to rural women by climate change, natural disasters, land and soil degradation, water pollution, droughts, floods, desertification, pesticides and agro-chemicals, extractive industries, monocultures, bio-piracy, and loss of biodiversity, particularly agro-biodiversity.'[223]

The human right to water as formally recognised in 2010 by the UN General Assembly differs from the right formulated in CEDAW. The general human right to water has been framed from a public health perspective and thereby prioritises water for domestic purposes. However, CEDAW recognises that especially for rural women water has importance for other purposes such as horticulture, crop irrigation and raising livestock. The CEDAW provision concentrates on the perspective of domestic-plus services which could benefit especially women.[224] It should be recognised that even in the absence of domestic-plus services, water users and especially women in this case will use the water they have access to in ways that they deem fit, thereby disregarding the artificial divide between water for domestic and other purposes.[225]

---

221 Ralph P Hall, Barbara Van Koppen and Emily Van Houweling, 'The Human Right to Water: The Importance of Domestic and Productive Water Rights' (2013) 20 Science and engineering ethics <www.ncbi.nlm.nih.gov/pubmed/24337891> accessed 25 July 2014. p. 849.

222 UN Committee on the Elimination of Discrimination Against Women, 'General Recommendation No. 34 on the Rights of Rural Women' 1. Paragraph 10.

223 Ibid. paragraph 12.

224 Hall, Van Koppen and Van Houweling. p. 850.

225 Gf White, D Bradley and Au White, 'Drawers of Water' (1972) 80 Bulletin of the World Health Organization 63 <www.who.int/entity/bulletin/archives/80(1)63.pdf>.

The right to water as derived from CEDAW stems from the right to adequate living standards. The article is not strongly formulated as it only asks for appropriate measures. In addition, the provision does not include a definition of its scope and does not allow for individual claims.[226] The obligation of states to give special attention to vulnerable groups also applies to involving women in procedural terms.[227] This is highly relevant as even within the same households, men and women have different interests which may conflict.[228] In addition, involving women may be difficult because of their household responsibilities in many societies.[229]

2.6.2 1989 CONVENTION ON THE RIGHTS OF THE CHILD

The Convention on the Rights of the Child was unanimously adopted by the UN General Assembly in 1989. According to the preamble to the CRC, children need special care and assistance. Hence, the Convention was adopted and includes special measures which States can take. The Convention explicitly mentions water as a part of the right to the highest attainable standard of health. Article 24(2)(c) of the 1986 Convention on the Rights of the Child (CRC) focuses on water as a means to combat disease and malnutrition. The article reads:

1. *States Parties recognize the right of the child to the enjoyment of the highest attainable standard of health and to facilities for the treatment of illness and rehabilitation of health. States Parties shall strive to ensure that no child is deprived of his or her right of access to such health care services.*
2. *States Parties shall pursue full implementation of this right and, in particular, shall take appropriate measures:*
   (c) To combat disease and malnutrition, including within the framework of primary health care, through, inter alia, the application of readily available technology and through the provision of adequate nutritious foods and ***clean drinking-water****, taking into consideration the dangers and risks of environmental pollution;*

The provision under the CRC covers several aspects of a human right to water. Firstly, it covers access to water for the preparation of food. Quantities of safe drinking water are needed for the preparation of significant amounts of foodstuffs such as infant

See Hall, Van Koppen and Van Houweling. on how water supplies designed for domestic use are used for productive purposes. See also Harris. p. 14 'Additionally, it was observed on several occasions during my fieldwork that women also use irrigation canalets for domestic uses, for instance, for dishes or laundry.'

226 Knut Bourquain, *Freshwater Access from a Human Rights Perspective* (Brill Academic Publishers 2008). p. 123.

227 Winkler, *The Human Right to Water Significance, Legal Status and Implications for Water Allocation*. p. 223.

228 United Nations, 'Women and Water' (2005). p. 4.

229 Filmer-Wilson. p. 235.

formula.[230] This holds especially true for liquid and semi-liquid nutrients such as dried milk. Dried milk is one of the most commonly used baby foods around the world.[231] Secondly, the provision covers a right to safe drinking water. Thirdly, it takes aspects of a healthy environment into account.[232]

The provisions of the CRC are not applicable to all individuals and therefore the conclusion of a human right to water based on its provisions cannot be generalised.[233] Nevertheless, the CRC emphasises the underlying determinants of health in more detail. The provision '*does not provide children with a generic right to freshwater access but merely specifies their right to health with regard to some positive obligations.*'[234] The reason behind this explicit mention of a human right to water could be the role that specific Conventions have in addressing and securing the rights of different vulnerable groups. The added value in the effectuation of these rights, however, still needs to be proven.[235]

2.6.3 2006 CONVENTION ON THE RIGHTS OF PERSONS WITH DISABILITIES

Persons with disabilities face major challenges in securing their access to water. However, clear data on the magnitude of the problem are lacking.[236] Facilities for water and sanitation are frequently not inclusive and this often exacerbates stigma and discrimination.[237] 'Social barriers vary in different cultures; however studies from many countries report that persons with disabilities often face stigma and discrimination from others when using both household and public facilities. Fears that persons with disabilities would contaminate water sources or that they would make the latrines dirty are frequently reported.'[238]

In order to protect and secure access to water for this vulnerable group Article 2 of the Convention on the Rights of Persons with Disabilities (CRPD) was adopted and it reads:

---

230 World Health Organization, *Protecting Groundwater for Health Managing the Quality of Drinking-Water Sources WHO Drinking-Water Quality Series Protecting Groundwater for Health MANAGING THE QUAALITY OF DRINKING-WATER SOURCES*. p. 99.

231 Thorsten Kiefer and Catherine Brölmann, 'Beyond State Sovereignty: The Human Right to Water' (2005) 5 Non-State Actors and International Law 183 <www.ingentaselect.com/rpsv/cgi-bin/cgi?ini=xref&body=linker&reqdoi=10.1163/157180705775435465>. p. 197.

232 Similar to Article 12 ICESCR.

233 Winkler, *The Human Right to Water*. p. 61.

234 Bourquain. p. 122.

235 Ibid. p. 123.

236 Jacqueline Noga and Gregor Wolbring, 'The Economic and Social Benefits and the Barriers of Providing People with Disabilities Accessible Clean Water and Sanitation' (2012) 4 Sustainability 3023. p. 3023 and Groce and others, 'Water and Sanitation Issues for Persons with Disabilities in Low- and Middle-Income Countries: A Literature Review and Discussion of Implications for Global Health and International Development.' p. 623.

237 World Health Organization, 'GLAAS 2012 Report. UN-Water Global Analysis and Assessment of Sanitatioin and Drinking-Water'. p. 47.

238 Groce and others, 'Water and Sanitation Issues for Persons with Disabilities in Low- and Middle-Income Countries: A Literature Review and Discussion of Implications for Global Health and International Development.' p. 619.

States Parties recognize the right of persons with disabilities to an adequate standard of living for themselves and their families, including adequate food, clothing and housing, and to the continuous improvement of living conditions, and shall take appropriate steps to safeguard and promote the realization of this right without discrimination on the basis of disability.
States Parties recognize the right of persons with disabilities to social protection and to the enjoyment of that right without discrimination on the basis of disability, and shall take appropriate steps to safeguard and promote the realization of this right, including measures:
(a) To ensure equal access by persons with disabilities to clean water services, and to ensure access to appropriate and affordable services, devices and other assistance for disability-related needs

This wording focusses on the barriers to accessibility. Similar to the ICESCR, the CRPD also includes a provision (Article 28) on an adequate standard of living.[239] The literature agreed that: 'barriers to access arise from obstacles in the built environment, social barriers and institutional factors, as well as individual abilities, and this range of barriers needs to be addressed holistically.'[240] However, the natural environment is often ignored.

2.6.4 INDIGENOUS PEOPLES

Indigenous peoples make up around five percent of the world's population, although they take up around 22 percent of the Earth's surface.[241] They are highly dependent on ecosystems as a source of livelihood and are thereby disproportionately affected by environmental degradation. Indigenous peoples can therefore also be considered to be a vulnerable group whose realisation of the right to water is fragile due to a combination of environmental and social factors. This group has a special relationship with the environment and with water as has been recognised in the 2003 Indigenous Peoples' Kyoto Water Declaration:

Our relationship with our lands, territories and water is the fundamental physical cultural and spiritual basis for our existence. This relationship to our Mother Earth requires us to conserve our freshwaters and oceans for the survival of present and future generations. We assert our role as caretakers with rights and responsibilities to defend and ensure the protection, availability and purity of water. We stand united to follow and implement our knowledge and traditional laws and exercise our right of self-determination to preserve water, and to preserve life.[242]

239 Doug Donoho, 'Some Critical Thinking about a Human Right to Water' (2012) 19 ILSA Journal of International and Comparative Law 91. p. 98.

240 N Groce and others, 'Water and Sanitation Issues for Persons with Disabilities in Low- and Middle-Income Countries: A Literature Review and Discussion of Implications for Global Health and International Development' (2011) 9 Journal of water and health 617. p. 620.

241 Mearns and Norton. p. 18 This 22 percent of the Earth's surface includes 80 percent of the remaining biodiversity and 90 percent of cultural diversity.

242 Http://portal.unesco.org/science/en/ev.php-URL_ID=3886&URL_DO=DO_TOPIC&URL_SECTION=201.html.

Contrary to the rights to water which have been formulated for children, rural women and the disabled, indigenous peoples currently do not have an explicit legally binding right to water which specifically applies to them. Nevertheless, there are provisions in international instruments which apply to or are relevant to indigenous peoples' access to water.

Besides the human right to water being derived from the ICESCR as articulated in General Comment No. 15, there are other provisions in international law from which rights to water can be derived, such as Article 6 ICCPR on the right to life. These other provisions can be invoked by individuals whether or not they are indigenous.[243] However, indigenous peoples have additional avenues for protecting their access to water. The self-determination of indigenous peoples and their participation in development are crucial aspects for protecting indigenous peoples' access to water. ILO Convention No. 169 includes requirements for states concerning consultation but does not include specific provisions on water resources. Those provisions can be found elsewhere. The following figure provides an overview of those provisions in international law which do specifically mention water resources.

Table 2.1 Overview of relevant provision for indigenous peoples' right to water

Instrument	Relevant provisions for indigenous peoples' right to water[244]
Convention on Biological Diversity (1992)*	- the objective is to sustain all life on Earth, including aquatic ecosystems, with the global goal being to reverse and stop the loss of biodiversity - provides for the respect, preservation and maintenance of knowledge, innovations and practices of indigenous and local communities relevant for the conservation and sustainable use of biological diversity - many of the decisions of the COP call for the full and effective participation of indigenous communities in order to achieve the global goal
Agenda 21 (1992)	- provides for the protection and management of freshwater resources recognising the effects that climate change will have on water and indigenous peoples. Identifies the need to: - engage indigenous peoples in water management policy-making and decision-making - improve indigenous technologies to fully utilise limited water resources and to safeguard those resources against pollution - recognise the interconnection between economic development and access to and the supply of water
United Nations Declaration on the Rights of Indigenous Peoples (2007)	- Indigenous access, conservation and economic development of water - a right to maintain and strengthen the distinctive indigenous spiritual relationship with 'traditionally owned or otherwise occupied and used lands, territories, waters and coastal seas' - the right to conservation and protection of indigenous lands and resources with state assistance - the right to development for all indigenous lands and resources including water.

243 Discussion in chapter 3.

244 The asterisk indicates the legally binding status of the instrument.

The above-mentioned provisions together with the general applicable human rights to water form a bundle of water rights which indigenous communities can invoke to protect not only their access to water for domestic purposes but to protect their access to water resources in general. Indigenous peoples can thereby use water resources in a way that they deem appropriate for cultural reasons, for subsistence agriculture, for environmental purposes and for other purposes. In the specific case of indigenous peoples there are often deep structural problems which need to be overcome such as a lack of infrastructure due to a lack of investment in minority-dominated regions and unjust land rights.[245] Also, tensions can occur as proposed state measures for mitigating climate change or environmental degradation can undermine the customary rights of indigenous peoples.[246]

2.6.5 FUTURE GENERATIONS

Although the majority of philosophers argue that future generations cannot have anything until the future has become the present, many lawyers are developing a sound reasoning for recognising the rights of future generations.[247] The rights of future generations have not been fully articulated but their needs are nevertheless sometimes mentioned in certain instruments. For instance, in Principle 3 of the Rio Declaration on Environment and Development.[248] The concept of sustainable development demonstrates this by expressing that the needs of the present generation must not comprise the ability of future generations to foresee their own needs.[249] Especially in relation to water resources which in many circumstances can be considered finite resources, taking them into consideration is highly relevant. General Comment No. 15 also recognises this and mentions that 'states parties should adopt comprehensive and integrated strategies and programmes to ensure that there is sufficient and safe water for present and future generations.'[250]

As mentioned before, it is expected that the number of individuals whose access to water is uncertain will increase. Therefore it can be assumed that if the necessary adjustments are put in place to provide access for those who are currently vulnerable this will positively affect future generations. For instance, the capital of Yemen, Sana'a,

245 Castellino. p. 15.

246 See for instance Ellen Desmet, 'The Twin of Green The Rights of Indigenous Peoples and Local Communities Entwined with Nature Conservation' (Katholieke Universiteit Leuven 2010) <https://lirias.kuleuven.be/handle/123456789/259914>.

247 See for instance Marcus Duwell, 'Human Dignity and Future Generations' in Marcus Duwell and others (eds), *The Cambridge Handbook of Human Dignity Interdisciplinary Perspectives* (Cambridge University Press 2014).

248 The right to development must be fulfilled so as to equitably meet developmental and environmental needs of present and future generations.

249 Brundtland Commission.

250 Committee on Economic Social and Cultural Rights. paragraph 28.

is expected to run out of water by 2025.[251] Yemen does not have permanent rivers and therefore it mostly depends on groundwater.[252] The over-extraction of this groundwater, for mostly agricultural purposes, has been taking place for several decades. As a result future generations will not be guaranteed access to water with all the resulting consequences. The consequences are already being felt as there are many internal refugees due to droughts in certain regions. In addition, many (armed) conflicts lie at the heart of water conflicts.[253] If the state fails to act now, this will lead to fewer and worse possibilities for future generations.[254]

2.7 CONCLUSION

The Committee on Economic, Social and Cultural Rights, which has authoritatively interpreted the human right to water, has adopted a limited approach to vulnerability which does not provide sufficient protection in relation to the human right to water. Just like other international human rights authorities, such as the European Court of Human Rights, it mainly considers social factors to be the causes of vulnerability. It does however implicitly include environmental factors, but does not do this in a consistent manner. In relation to the topic of water management, this categorisation is insufficient. For this reason other factors should be included.

Social vulnerability does not sufficiently recognise the situation that vulnerable groups are in as most people who currently lack access to water currently live in unserved and underserved areas where they depend on traditional sources of water. Ensuring that their environment has sufficient and clean water available will allow them to facilitate their own human right to water. In addition, the current limited scope of vulnerability means that other vulnerable groups which are in need of special protection, but do not mirror the prohibited ground of discrimination, are not included within the ambit of this protection.

Vulnerability not only depends on the status of the group in question according to social, cultural and political factors or inherent conditions such as gender, but also the specific situation they are in, in a specific place/environment and even at a specific moment. Vulnerability is a dynamic condition in which people move back and forth between vulnerable and secure, depending on their capacity to adjust.[255] For instance, a community which is already in a situation of poverty and has limited coping capacity

---

251 Christopher Boucek, 'Yemen: Avoiding a Downward Spiral' (2009) 102. p. 6.

252 Taha Taher and others, 'Informing and Enabling Local Ground Water Governance for Yemen' (2011). p. 9.

253 See Small Arms Survey, 'Under Pressure: Social Violence over Land and Water in Yemen' (2010) <www.genevadeclaration.org/fileadmin/docs/regional-publications/Social_violence_over_land_and_water_in_Yemen.pdf>.

254 Boucek. p. 4.

255 Mike Brklacich, May Chazan and Hans-Georg Bohle, 'Human Security, Vulnerability, and Global Environmental Change' in Richard A Matthew and others (eds), *Global Environmental Change and Human Security* (The MIT Press 2010). p. 37.

will be pushed into intensified vulnerability by even relatively modest environmental exposure or deterioration.[256]

Adding environmental factors is not an unwarranted expansion of human rights law. The environment is a prerequisite for enjoying human rights and this implies that states have a duty to ensure a quality of the environment which allows for the full exercise of protected rights.[257] In addition, a trend can be noticed where environmental concerns are influencing human rights law. If international human rights law aims to effectively take on issues concerning the environment, such as climate change and environmental degradation, the environmental vulnerability of individuals and groups should be part of its considerations. In addition, enhancing human rights will reduce the environmental vulnerability of certain groups.[258]

The added value of classifying certain groups as vulnerable is that their specific needs will be taken into account during implementation and will thereby make the realisation of the right more effective. It is becoming increasingly clear that ecological and geographical factors are not disconnected from the social and institutional context.[259] One of the best approaches that states can take for protecting human rights is to establish processes which ensure that violations will not take place.[260] This is particularly important in those circumstances where remedying negative impacts is difficult if not impossible, which is often the case in relation to water resources such as groundwater.

The situation of vulnerable groups is also connected to sustainability. Overlooking the needs and interests of a section of the population is thought to undermine the sustainability of the services provided.[261] Several studies have shown that the inclusion of different marginalised groups in WASH projects supports the longevity of these projects.[262] Also, the most vulnerable tend to contribute the least to environmental degradation and can also exert the least influence on regulatory policies.[263]

As mentioned, rights to water are already recognised for different vulnerable groups. However, the approach of these provisions is limited as it deals with the identity of vulnerable groups and individuals in a one-dimensional fashion. Also, the instruments have limited measures for enforcement. Including social as well as environmental factors will allow the interests of more relevant vulnerable groups to be addressed by specific measures which will support the realisation of their human right to water.

256 Ibid. p. 38.

257 Dinah Shelton, 'Human Rights, Environmental Rights, and the Right to Environment' (1991) 28 Stanford Journal of International Law 103. p. 103.

258 Barnett. p. 265.

259 Brklacich, Chazan and Bohle. p. 41.

260 Mearns and Norton. p. 14.

261 Harris. p. 2.

262 See for instance Susan Margaret Roelofs Greenfield, 'Challenging the Rhetoric of Sustainability and Participation a Case Study of the Ghana Upper Region Water Supply Programme.pdf' (Carleton University 1990). and Nyree Stenekes, 'Sustainability and Participation in the Governing of Water Use: The Case of Urban Water Recycling' (University of New South Wales 2006). p. 126.

263 Brklacich, Chazan and Bohle. p. 46. This ties into the principle of common but differentiated responsibilities and intergenerational equity on the local level.

CHAPTER 3
THE HUMAN RIGHT TO WATER IN INTERNATIONAL HUMAN RIGHTS LAW

"Access to safe water is a fundamental human need and, therefore, a basic human right. Contaminated water jeopardizes both the physical and social health of all people. It is an affront to human dignity."[264]

3.1 INTRODUCTION

Chapter 1 gave an introduction into the subject of this book. As explained, the focus of this research is on the realisation of vulnerable groups' sustainable human right to water. It introduced the topic and some of the bottlenecks regarding the human right to water and sustainability. Chapter 2 argued that the concept of vulnerability within the human rights framework is currently too narrow to facilitate additional support and access to water for those groups who can be identified as specifically vulnerable in relation to water and other environmental human rights. It argued that environmental factors should be taken into account in assessing the vulnerability of individuals and groups[265], thereby recognising the impact that their situation has on the realisation of their rights.

The recognition of the human right to water provides an opportunity for these groups to have their access to water realised. In order to clarify which rights individuals have in general and vulnerable groups in particular, this chapter discusses what the human right to water entails in international human rights law. This includes its legal foundations and normative content. The existence of the human right to water has been contested, however. With the adoption of General Comment No. 15 by the UN Committee on Economic, Social and Cultural Rights (CESCR), however, the right to

264 Kofi Annan, former United Nations Secretary-General.

265 See Chapter 2.

water and sanitation[266] have been authoritatively formulated in a non-legally binding instrument.[267]

This chapter commences with an overview of how the human right to water has emerged in the international arena and which developments it has experienced. This section illustrates that access to water was initially understood as a collective need, taking into account environmental protection, and it developed into a human right with an individual scope. It has subsequently moved back and forth between the environmental policy field and the human rights field which is a relevant development in relation to sustainable development.[268] This cross-fertilisation and how it relates to aligning the human right to water and sustainability efforts will be further discussed in chapter 5.

After discussing its emergence and evolution, this chapter explores the legal foundations of the human right to water. It is often assumed that there is a human right to water. However, in order to make such a claim it is required to substantiate this notion.[269] It is necessary to explain in which manner this right exists and on which legal basis. In international human rights law different provisions can be found on which the human right to water can be based. There are several recognised human rights of which access to water is an essential element.[270] It is imperative to discuss the foundation of the right as it determines the characteristics of the right and how it should be implemented. It also assists in laying the basis for understanding the place of the human right to water in overall water management.

The human right to water is still at the development stage at the international level. However, developments at regional and national level are taking place which inform how the right can be interpreted and implemented.[271] Several national constitutions and lower legislative instruments include fundamental rights from which rights to water can

266 Although the human right to water and the right to sanitation are often seen as being coupled together, this book focusses and discusses the human right to water. See Keri Ellis and Loretta Feris, 'The Right to Sanitation: Time to Delink from the Right to Water' (2014) 36 Human Rights Quarterly 607 <http://muse.jhu.edu/content/crossref/journals/human_rights_quarterly/v036/36.3.ellis.html> accessed 11 August 2014., Pedi Obani and Joyeeta Gupta, 'The Evolution of the Right to Water and Sanitation: Differentiating the Implications' (2015) 24 Review of European, Comparative and International Environmental Law 27. and (Winkler, 2016) on the right to sanitation and its evolution. Section … discusses the right to sanitation in relation to the right to water.

267 Committee on Economic Social and Cultural Rights, 'General Comment No. 15 The Right to Water' (2002).

268 See section 2.

269 Stephen Tully, 'A Human Right to Access Water? A Critique of General Comment No. 15' (2005) 23 Netherlands Quarterly of Human Rights 35. p.35.

270 See section 3.

271 'Drawing upon sources found in international law, not as binding rules but as contextual principles, judges of municipal courts in this century will assume an important function in making the principles of international law a reality throughout the world. We cannot leave this function to international courts and tribunals alone.' Michael Kirby, 'International Law – The Impact on National Constitutions' (2006) 21 American University International Law Review 327. p. 329.

be derived.[272] Analysing how the human right to water has been derived and interpreted at the national level can assist in understanding the human right to water at the international level.[273] For this reason, this chapter includes, where possible, examples of national provisions as illustrations of the different legal foundations of the right and how these different foundations affect the definitions of the human right to water.

After having established the legal foundations of the human right to water, it will be briefly considered how access to water for domestic purposes can be prioritised. After this the content of the right will be discussed as it is currently understood according to the most authoritative interpretation of the right in General Comment No. 15 of the Committee on Economic, Social and Cultural Rights (CESCR).[274] That section will highlight, among other things, the aspects of water use covered by the human right to water. The content will make clear which entitlements individuals have. Here it will also be discussed which general and specific obligations states have regarding this right. The chapter will be concluded in section 3.6.

3.2 EMERGENCE OF THE HUMAN RIGHT TO WATER

Access to water as a human right to water has been highlighted at different international forums over the years. This section provides an overview of the significant developments surrounding this topic. The concept of a human right to water as a recognised human right is relatively new within the international human rights law arena. This right was not originally included in the catalogue of international human rights.[275] Access to water first surfaced in international environmental policies before being included within the field of human rights. Access to water is of course not only a human rights concern but is also important in other fields like economics and governance.[276] The developments surrounding entitlements regarding access to water highlight the different priorities that access to water has had and the perspectives from which states looked at the issue throughout the previous decades.[277] Different values

---

272 South Africa and India are often referred to as the most progressive examples of the national implementation of the human right to water. See Thorsten Kiefer and others, *Legal Resources for the Right to Water and Sanitation* (2nd edn, Centre on Housing Rights and Evictions 2008) <www.cohre.org/water>. for selected constitutional sources and other national laws and policies.

273 Obani and Gupta, 'The Evolution of the Right to Water and Sanitation: Differentiating the Implications'. p. 34.

274 Takele Soboka Bulto, 'The Emergence of the Human Right to Water in International Human Rights Law : Invention or Discovery?' (2011) 7.

275 Tully. p. 36.

276 Malcolm Langford, 'The United Nations Concept of Water as a Human Right: A New Paradigm for Old Problems?' (2005) 21 International Journal of Water Resources Development 273 <www.tandfonline.com/doi/abs/10.1080/07900620500035887> accessed 6 March 2013. p. 274. The approaches are not mutually exclusive.

277 See for instance principle 4 of the Dublin Statement on Water and Sustainable Development: 'Managing water as an economic good is an important way of achieving efficient and equitable use, and of encouraging conservation and protection of water resources.'

such as economic value, environmental protection and equal access, relevant for access to water, have occupied centre stage at different times.[278] The following section shows these developments and how the human right to water has morphed into its current articulation and which characteristics it has adopted.

In 1972, the debate on the right to water within the international arena started with the Stockholm Declaration adopted at the UN Conference on the Human Environment.[279] Here it was agreed that water as a natural resource should be safeguarded.[280] It was declared that:

> *'The natural resources of the earth including the air, water, land, flora and fauna and especially representative samples of natural ecosystems must be safeguarded for the benefit of the present and future generations through careful planning or management, as appropriate.'*[281]

The start of the water access debate therefore originated with a sustainability point of view where benefits for future generations were taken into account. However, the provision is not legally binding for states and does not establish entitlements for individuals.

During the 1990 Global Consultation on Safe Water and Sanitation the New Delhi Statement was formulated, summarising conclusions and recommendations for realising universal water supply and sanitation.[282] The New Delhi Statement, adopted by 115 states[283], recognises several issues concerning water resources, such as depletion and degradation, and encourages states to take radical approaches to evade water crises.[284] A right to water has not been articulated in this document. It 'formalized the need to provide, on a sustainable basis, access to safe water in sufficient quantities and proper sanitation for all.'[285]

The Stockholm Declaration was followed by the 1977 UN Water Conference in Mar del Plata.[286] The Mara del Plata Action Plan, adopted during this conference, deals

---

278 See Joseph W Dellapenna and Joyeeta Gupta, 'The Evolution of Global Water Law' in Joseph W. Dellapenna; Joyeeta Gupta; (ed), *The Evolution of the Law and Politics of Water* (2009th edn, Springer Netherlands 2009).

279 Declaration of the United Nations Conference on Human Environment, available on: www.un-documents.net/unchedec.htm last visited: May, 2017.

280 Declaration of the United Nations Conference on Human Environment, June 16, 1972, U.N. Doc. A/ CONF.48/14/Rev.1, Sales No.E. 73.II.A.14 (1973) principle 2. See also Salman M. A. Salman; Siobhán McInerney-Lankford, *The Human Right to Water* (David Freestone; Salman MA Salman; ed, 2004th edn, The International Bank for Reconstruction and Development / The World Bank 2004). p. 7.

281 Principle 2 of the Stockholm Declaration.

282 See Martin G Beyer, 'The Global Consultation on Safe Water and Sanitation for the 1990s' (1991) 15 Natural Resources Forum 118.

283 WHO/UNICEF, '25 Years Progress on Sanitation and Drinking Water' (2015). p.30.

284 Gonzalo Aguilar Cavallo, 'The Human Right to Water and Sanitation : From Political Commitments to Customary Rule ?' (2012) 3 Pace International Law Review Online Companion 136. p. 159.

285 Agenda 21 1992 1. paragraph 18.48.

286 Report of the United Nations Water Conference, Mar del Plata, 14–25 March 1977 available on: www.internationalwaterlaw.org/bibliography/UN/Mar_del_Plata_Report.pdf last visited in December, 2012.

with major issues concerning water resources such as water management and pollution control. The Declaration mentions that 'all peoples whatever their stage of development and their social and economic conditions, have the right to have access to drinking water in quantities and of a quality equal to their basic needs.'[287] This statement, unlike the Stockholm Declaration, has been formulated in a more rights-based language. Even though it is not legally binding, this statement can be seen as the genesis of the human right to water. After this initiative, others, such as the 1992 Dublin Statement, supported this rights-based approach by, for instance, recognising water as an economic good without overlooking its role as a 'basic right of all human beings to have access to clean water and sanitation at an affordable price.'[288] Nevertheless, also this provision cannot be used by individuals when claiming an entitlement.

The 1992 Dublin Statement shows that the economic value of water may come into competition with other values such as access to water for vulnerable groups and environmental protection.[289] The global free-market economy and state sovereignty policies risk excluding the influence of vulnerable individuals if they lack economic influence.[290] 'States have historically not provided water to many poor and marginalised communities, forcing them to purchase water from street vendors or tanker trucks at prices inflated up to twelve times the price of water from municipal supply systems.'[291] In the eyes of many, the Dublin Declaration fell short, resulting in a limited formulation of this right.[292] And this while the privatisation of public water services was increasing in many developing countries such as Bolivia where communities took to the streets in order to ensure that their access to water remained financially accessible.[293]

In Chapter 18 of Agenda 21, formulated at the 1992 United Nations Conference on Environment & Development[294],on the protection of the quality and supply of freshwater

---

287 Report of the United Nations Water Conference, Mar del Plata, 14–25 March 1977 (United Nations publication, Sales Np. E.77.II.A.12), part 1, chapter I, section C, para. 35. See Knut Bourquain, *Freshwater Access from a Human Rights Perspective* (Brill Academic Publishers 2008). p. 12.

288 Salman 2004, p. 9 Dublin Statement on Water and Sustainable Development, see Journal of Water SRT Aqua, Vol. 41, No. 3, at 129.

289 International Conference on Water and Environment, Dublin, Ir., Jan. 26–31, 1992, The Dublin Statement on Water and Sustainable Development, at 7 (1992). There are three features connected to the commodity approach: water should be priced, water delivery should be opened to the market and water resources should be privately owned. Langford 2005, p. 274.

290 Erik B Bluemel, 'The Implications of Formulating a Human Right to Water' (2004) 31 Ecology Law Quarterly 957 <http://heinonlinebackup.com/hol-cgi-bin/get_pdf.cgi?handle=hein.journals/eclawq31&section=42>. p. 963.

291 Ibid. p. 959.

292 Ibid., p. 967.

293 Shawkat Alam, Natalie Klein and Juliette Overland, *Globalisation and the Quest for Social and Environmental Justice* (2010). p. 62 Also see Oriol Mirosa and Leila M Harris, 'Human Right to Water: Contemporary Challenges and Contours of a Global Debate' (2012) 44 Antipode 932 <http://doi.wiley.com/10.1111/j.1467-8330.2011.00929.x> accessed 6 March 2013. p. 935 on large-scale protests against water privatization.

294 UNCED 1992 Agenda 21, Chapter 18. Protection of the Quality and Supply of Freshwater Resources: Application of Integrated Approaches to the Development, Management and Use of Water Resources. United Nations Conference on Environment and Development, Rio de Janeiro, 14 June 1992.

resources recommendations on sound water resources management are laid down. The document recognises basic human needs. However, it does not formulate a right which can be enforced but places the basic needs of humans in the bigger picture of water management. It for instance states that: 'in developing and using water resources, priority has to be given to the satisfaction of basic needs and the safeguarding of ecosystems. Beyond these requirements, however, water users should be charged appropriately.'[295]

In 1997, the UN Sub-Commission on the Protection and Promotion of Human Rights requested Mr El-Hadji Guissé to explore the topic of the right of access to drinking water and sanitation.[296] This constituted the first step towards its recognition as a human right. In 2002, the Committee on Economic, Social and Cultural Rights took the lead in the human right to water debate by adopting General Comment No. 15 on the human right to water.[297] However, General Comment No. 15 is not legally binding. Nevertheless, it is seen as the most authoritative interpretation of the right to water. It establishes the right as being self-standing while it is also essential for the right to an adequate standard of living and is related to the right to the highest attainable standard of health and the rights to adequate housing and adequate food.[298] This articulation of the right includes several water uses such as domestic purposes and personal hygiene. However, it seems to disregard environmental issues as a central element, while they were indeed included when the right was still essential in the field of environmental policy. Even with the creation of General Comment No. 15 scepticism still remained as mostly Western states questioned its added value, its legal status and legal foundation.[299]

Nonetheless, with the adoption of Resolution 64/292 in 2010, the UNGA explicitly recognised 'the right to safe and clean drinking water and sanitation as a human right that is essential for the full enjoyment of life and all human rights.'[300] After the General Assembly the Human Rights Council adopted a resolution which states that: 'the human right to safe drinking water and sanitation is derived from the right to an adequate standard of living and inextricably related to the right to the highest attainable standard of physical and mental health, as well as the right to life and human dignity.' In 2014 the Human Rights Council reaffirmed its previous resolution and stated 'that safe drinking water and sanitation must be progressively made available for present and future generations, without discrimination, and that the provision of services today

---

295 Chapter 18.8.

296 In 1997 the UN Sub-Commission on the Protection and Promotion of Human Rights requested Mr El-Hadji Guissé to explore the topic of the right of access to drinking water and sanitation. Sub-Commission on the Promotion and Protection of Human Rights (Sub-Commission) Resolution No. 1997/18Mr. El-Hadji Guissé was subsequently appointed as the Special Rapporteur on the topic. He presented his report in 2005. U.N. Doc. E/CN.4/Sub.2/2005/25, Sub-Commission for the Promotion and Protection of Human Rights, Realization of the right to drinking water and sanitation, report of the Special Rapporteur, El Hadji Guissé, July 2005.

297 Committee on Economic, Social and Cultural Rights, 29th session, 20 January 2003, E/C.12/2002/11 www.unhchr.ch/tbs/doc.nsf/0/a5458d1d1bbd713fc1256cc400389e94/$FILE/G0340229.pdf.

298 Ibid.

299 Doug Donoho, 'Some Critical Thinking about a Human Right to Water' (2012) 19 ILSA Journal of International and Comparative Law 91. p. 101 Engburch 2006, p. 129 – 130.

300 Resolution A/RES/64/292. United Nations General Assembly, July 2010.

should safeguard the ability in the future to realise the human right to safe drinking water and sanitation.'[301]

In 2015, the 2030 Agenda for Sustainable Development, often referred to as the Sustainable Development Goals (SDGs)[302], was adopted following the Millennium Development Goals (MDGs) which had already been concluded.[303] Goal six is to 'ensure availability and sustainable management of water and sanitation for all' and is accompanied by several targets to measure progress. Once again, access to water was not approached as a right. However, it does place access to water for individuals in the larger picture of water management.[304]

Table 3.1 Overview elements in water access provisions

Year	Instrument[305]	Right	Quality	Quantity	Access	Clean water recourses
1977	Mara del Plata Declaration	x	x	x	x	
1990	New Delhi Statement		x		x	x
1992	Dublin Statement on Water and Sustainable Development	x	x		x	x
1992	Chapter 18 Agenda 21		x	x	x	x
1999	Protocol on Water and Health*		x	x	x	x
2001	Bonn International Conference on Freshwater		x	x	x	x
2002	Johannesburg Report and Plan of Implementation		x		x	x
2002	General Comment No. 15	x	x	x	x	x
2004	ILA Report		x	x	x	x
2005	UN Millennium Development Goals		x		x	x
2010	UNGA Resolution 64/292 on the Right to Water and Sanitation	x	x		x	
2010	UNHRC Resolution 18/1 on Human Rights and Access to Safe Drinking Water and Sanitation	x	x	x	x	
2014	UNHRC Resolution 27/ 7 on the Rights to Water and Sanitation	x	x	x	x	
2015	UN Sustainable Development Goals		x	x	x	x

301 Paragraph 2 27/7, The human right to safe drinking water and sanitation.

302 UN General Assembly: Transforming Our World: The 2030 Agenda for Sustainable Development. UN Doc; A/RES/70/1, 21 October: 2015.

303 UN General Assembly: Transforming Our World: The 2030 Agenda for Sustainable Development. UN Doc; A/RES/70/1, 21 October: 2015.

304 See Goal 6: Ensure availability and sustainable management of water and sanitation for all. This goal includes for instance target 6.4: By 2030, substantially increase water-use efficiency across all sectors and ensure sustainable withdrawals and supply of freshwater to address water scarcity and substantially reduce the number of people suffering from water scarcity.

305 The asterisk indicated the legally binding nature of the instrument.

The overview shows the developments in access to water since the 1970s. The different elements demonstrate the different perspectives and focus on the instruments which deal with this topic. There seems to be a cross-fertilisation between international environmental policy and human rights. The overview does not however show how strongly the instruments emphasise these perspectives. The instruments also show that the awareness concerning the place of the human right to water in general water management is increasing.

3.3 LEGAL FOUNDATIONS OF THE HUMAN RIGHT TO WATER

As mentioned, the human right to water is not included in the catalogue of human rights. For this reason its existence has been questioned. To determine whether there is such a right, its legal foundation in the recognised sources of international law needs to be found. According to Article 38 of the Statute of the International Court of Justice, the accepted sources of international law are: 'international conventions, international custom, general principles of law and subsidiary means for the determination of rules of law such as judicial decisions and teachings of the most highly qualified publicists.'[306] This section looks at these different sources and analyses whether they provide a legal foundation for a human right to water. It starts off with the legal foundations in the ICCPR, then discusses the ICESCR. After this the possibilities for the human right to water to be a part of customary law and general principles of law are analysed.

The legal basis of the human right to water is essential for understanding which rights and obligations are created by the human right to water, and for whom and how the right will be enforced.[307] For instance, if the human right to water finds its legal basis in a treaty, those states which are parties to that treaty are bound by the obligations it creates.[308] If, however, the human right to water is found to be part of customary law or is recognised as a general principle, then the right creates obligations for all states.[309] The legal foundation of the right helps to answer these questions. It also answers the question of whether the human right to water is an independent or subordinate right. This is relevant because 'given the structure of international law, State obligations depend upon which human right a right to water is found to support or whether such a human right to water is ultimately found to be a separate and independent human right from other recognised human rights. Whether a human right to water is ultimately

306 Article 38 of Rome Statute A/CONF.183/9 of 17 July 1998 available on: www.icc-cpi.int/nr/rdonlyres/add16852-aee9-4757-abe7-9cdc7cf02886/283503/romestatuteng1.pdf.

307 Amanda Cahill, '"The Human Right to Water – a Right of Unique Status": The Legal Status and Normative Content of the Right to Water' (2005) 9 The International Journal of Human Rights 389 <www.tandfonline.com/doi/abs/10.1080/13642980500170840> accessed 26 December 2013. p. 398.

308 Rebecca Bates, 'The Road to the Well: An Evaluation of the Customary Right to Water' (2010) 19 Review of European Community and International Environmental Law 282.p. 283.

309 Ibid. p. 284.

established as a right subordinate to other human rights or as an independent human right, recognition of a human right to water will have far-reaching effects.'[310]

As discussed in the previous section, different instruments within public international law, specifically international human rights and environmental law, deal with access to freshwater. These instruments assist in shining light on what the right entails. Table 3.1 provides an overview of the elements that these instruments assign to accessing water. The instruments discussed in the following section can mainly be identified as traditional human rights instruments. In particular, the ICCPR and the ICESCR are strong contenders for providing a legal basis for the human right to water, due to the legal status of the Covenants and the wide support given to these treaties by states.[311] Especially General Comment No. 15 formulated by the ESCR Committee plays a central role in recognising the human right to water under the ICESCR. But also other relevant conventions and treaties assist in providing a legal basis.[312]

As mentioned above, the human right to water is not explicitly mentioned in the authoritative human rights treaties, neither in the ICESCR nor in the ICCPR. This does not mean, however, that there is no place for the human right to water within the framework of these instruments. A way of accomplishing this is by implicit interpretation.[313] What should be kept in mind is that placing the human right to water under either one or both of these instruments will influence the implementation, enforcement and content of the human right to water. A solution can be found by taking a different route. The following section therefore looks at the implicit recognition of the right to water under the provisions of both Covenants. In addition, this section discusses the explicit recognition of the right under other relevant international human rights treaties.

3.3.1 LEGAL FOUNDATIONS OF THE HUMAN RIGHT TO WATER IN THE 1966 INTERNATIONAL COVENANT ON CIVIL AND POLITICAL RIGHTS

The ICCPR was drafted together with the ICESCR in the 1960s. These two documents together form an authoritative catalogue of international human rights.[314] Seeking protection for the right to water under the ICCPR has several benefits. Firstly, civil and political rights still remain the more politically accepted catalogue of human

310 Bluemel. p. 957.

311 Eibe Riedel, Gilles Giacca and Christophe Golay, 'The Development of Economic, Social, and Cultural Rights in International Law', *Economic, Social, and Cultural Rights in International Law* (2014).p. 6.

312 As mentioned in chapter 2 rights to water have been included in the CEDAW, CRC and CRDP.

313 Stephen C McCaffrey, 'A Human Right to Water : Domestic and International Implication' (1992) 5 Georgetown International Environmental Law Review 1. p. 8.

314 Eibe Riedel, Gilles Giacca and Christophe Golay, *Economic, Social, and Cultural Rights in International Law* (Eibe Riedel, Gilles Giacca and Christophe Golay eds, Oxford University Press 2014). p. 6 – 8.

rights.[315] Therefore framing the human right to water as deriving from the rights formulated in the ICCPR will assist in accepting its legal nature. Secondly, more states are parties to the ICCPR than the ICESCR.[316] Countries such as the United States of America are only a party to the ICCPR and not the ICESCR. Treaty provisions only bind those who are parties and therefore having a greater number of parties is beneficial for the recognition of the right and creates entitlements for more people. Thirdly, the ICCPR has an enforcement mechanism, thereby allowing for the right to be protected at the international level. This enforcement mechanism allows cases to be reported to the Human Rights Committee[317] and includes the possibility to hear inter-state complaints.[318] This Protocol also provides the possibility for individuals to complain about infringements by states which are parties to the Protocol.[319] And the provisions of the ICCPR are interpreted as having immediate effect.[320]

The right to life (Article 6 ICCPR)

In general it can be said that the right to water contributes to the realisation of all the rights mentioned in the ICCPR[321], as all human rights are interlinked.[322] There is no explicit mention of the right to water in the Covenant, nevertheless there are two provisions from which the right to water can be implicitly derived namely Article 6 on the right to life and Article 7 on the right to be free from cruel, inhuman or degrading treatment.[323] Both of these provisions have also been used in national contexts for claiming protection of the human right to water for individuals. These examples are mentioned below.

The right to life is one of the first rights mentioned in the Universal Declaration on Human Rights[324] and has been included in the ICCPR. Article 6(1) explains that: 'Every human being has the inherent right to life. This right shall be protected by law. No one shall be arbitrarily deprived of his life.' Water is essential to sustain life. To deny access to water completely, resulting in death, is an absolute infringement of the right

315 P Thielborger, 'Re-Conceptualizing the Human Right to Water: A Pledge for a Hybrid Approach' (2015) 15 Human Rights Law Review 225 <http://hrlr.oxfordjournals.org/cgi/doi/10.1093/hrlr/ngv008>. p. 243.

316 The Covenant has a high number of ratifications, with 114 states parties.

317 Art. 40 ICCPR.

318 Art. 41–43 ICCPR.

319 Article 2 on communications see www.ohchr.org/Documents/Press/OP_ICESCR.pdf.

320 General Comment No. 31 of the Human Rights Committee: 'The article 2, paragraph 1, obligation to respect and ensure the rights recognized in the Covenant has immediate effect for all States parties. Article 2, paragraph 2, provides the overarching framework within which the rights specified in the Covenant are to be promoted and protected.' par. 5; McCaffrey. p. 9.

321 The Covenant has a high number of ratifications, with 114 parties, as of October, 2012.

322 Riedel, Giacca and Golay, 'The Development of Economic, Social, and Cultural Rights in International Law'. p. 34.

323 Article 7 ICCPR is also an option, but is very limited as it cannot appropriately serve as a provision providing individual rights to water. Read Bourquain. p. 134 – 135.

324 Article 3 UDHR.

to life.[325] Therefore the right to water forms an element of the right to life; however, the boundaries of a human right to water deriving from Article 6 ICCPR requires analysis.

This legal foundation of the right to water mostly creates negative obligations for the state to abide by: the prohibition of interference.[326] Article 6(1) ICCPR obliges states to respect the right to life, meaning that the state should refrain from actions which result in the death of right holders.[327] In relation to water, this means that states should not deprive the holders of the right to water to the extent that they will lose their lives as a consequence of dehydration or pollution. States should not interfere with the availability of freshwater in such a manner that it affects the right to life.[328] This requirement does not only concern the quantity of water available but also its quality.[329] It can also (for instance) be inferred that the right to water under Article 6 ICCPR has been infringed in cases where States are responsible for polluting water resources which has led to the deaths of citizens. And also if, for instance, state-supplied drinking water is of such a low standard that it endangers peoples' lives.[330] The question remains whether Article 6 (1) ICCPR, besides these negative obligations, also covers positive obligations for states to take active steps in realising the right to life by supplying life-sustaining access to water.[331]

The Human Rights Committee has stated in General Comment No. 6 that the rights formulated in the ICCPR should not be interpreted narrowly, meaning that the right to life does not merely include negative obligations where states only have the obligation not to deprive citizens of their lives.[332] General Comment No. 6 further specifies that the right to life 'cannot properly be understood in a restrictive manner, and the protection of this right requires that states adopt positive measures.'[333] The Committee considers it desirable for States to take active measures to eliminate malnutrition and epidemics as far as this helps to decrease infant mortality and increase life expectancy.[334] This interpretation allows one to contend that states are obliged to provide life-sustaining resources of freshwater for their citizens. Nevertheless, an analysis 'of legal doctrine

325 Thielborger 232.
326 Cahill. p. 397.
327 The arbitrary deprivation of life is especially prohibited under Article 6. See General Comment No. 6: Article 6 (Right to Life) 1982 1.
328 Bourquain., p. 125.
329 Ibid. p. 125.
330 See for instance Red de Vigilancia y Exigibilidad de los Derechos Económicos, Sociales y Culturales Región Junín c/ Municipalidad Provincial de Huancayo.
'Municipal authorities must comply with their obligation to protect the life, health and the environment of a population exposed to contaminated water provided by a public water company, under Peruvian law. Accordingly, the population must be informed of the risk to life and health caused by the delivery of such contaminated water' Waterlex; and WASH United, 'The Human Rights to Water and Sanitation in Courts Worldwide' (2014). p. 146.
331 Thorsten Kiefer and Catherine Brölmann, 'Beyond State Sovereignty: The Human Right to Water' (2005) 5 Non-State Actors and International Law 183 <www.ingentaselect.com/rpsv/cgi-bin/cgi?ini=xref&body=linker&reqdoi=10.1163/157180705775435465>. p. 185; Bluemel. p. 968.
332 General Comment No. 6: Article 6 (Right to Life). par. 5.
333 Committee on Economic Social and Cultural Rights. paragraph 5.
334 Ibid. paragraph 5.

and practice before the Human Rights Committee has produced very little conclusive support for the submission'[335] that the right to water can be derived from the positive obligations resulting from the right to life.

Cahill argues that the right to life is not understood as protection from the absence of requirements for minimum survival.[336] In this line of reasoning the absence of water to sustain life cannot be seen as an infringement of the right to water as interpreted under the right to life.[337] However, a deliberate denial of, for instance, food could be understood as such an infringement. McCaffrey, on the other hand, finds that Article 6 ICCPR indeed obligates states to provide sufficient water to sustain life. His argument is based on the general obligation of the ICCPR to impose the immediate obligation to respect and ensure the right to life.[338]

What can so far be discerned is that from the right to life as formulated in Article 6(1) ICCPR a human right to water can only be derived in so far as this extends to protecting the availability of water and its necessary quality in order to sustain human life.[339] In the academic literature the legal basis of the human right to water is mostly associated with Article 11 ICESCR: the right to an adequate standard of living. However, the significance of the right to life in relation to the right to water should not be overlooked[340], especially when a large number of states recognise that this right exists. Also, Article 4(2) ICCPR explains that Article 6 is one of the provisions from which states may not derogate. In times of emergency states may not suspend their efforts to realise the full enjoyment of this right.[341] Article 6 is considered by many scholars to be part of customary law, while some argue that this right has become a norm of jus cogens.[342] This means that it binds all states.

Next to these benefits there are several limitations to deriving the human right to water from Article 6 ICCPR. Arguments can be found to support the idea that Article 6 allows for a claim by individual water users in as far as this is meant to protect human life. In the situation where people have sufficient water to sustain life, the human right to life is not threatened or violated.[343] The reasons for limiting the right to water under Article 6 to a negative right with a very limited positive dimension can be found in the uncertainty that might be created by its scope and the nature of the states' obligations.[344] Also, the right to water might create a challenge for the justiciability of

---

335 Kiefer and Brölmann. p. 189.

336 Cahill. p. 397.

337 See Kiefer and Brölmann. p. 190 and Yoram Dinstein, 'The Right to Life, Physical Integrity, and Liberty' in Louis Henkin (ed), *The International Bill of Rights. The Covenant on Civil and Political Rights* (Columbia University Press 1981). p.115.

338 McCaffrey. p. 9.

339 Thielborger. p. 232.

340 Ibid. p. 232.

341 Art. 4(1) ICCPR.

342 Kiefer and Brölmann. p. 186.
Many scholars are of the opinion that some of the obligations found in the ICCPR are part of customary international law. (Henkin, 1995).

343 Ibid. p. 133.

344 Ibid. p. 189.

the right at the domestic level.[345] On top of this, the Human Rights Committee might encounter challenges in relation to the implementation of the right and the examination of state practice with regard to individual complaint procedures.[346]

In addition, widening the interpretation of the right to life to include the right to water could interfere with the interpretation of the right to water under the ICESCR.[347] The provisions under that Covenant go beyond the sustainability of life and negative obligations. The human right to water as explained by General Comment No.15 aims to facilitate human dignity, health and life. It is important to note that there is a difference in fulfilling the right as supporting human dignity and fulfilling the minimum core which is geared towards survival.[348] Therefore it can be said that in relation to Article 6(1) ICCPR the right to water does not go far beyond the negative aspect where water is used as a means to arbitrarily deprive the lives of citizens[349] and that this right only extends to providing the amount of water necessary to redress the threat to life and therefore excludes uses such as personal and domestic uses which go beyond this.[350]

An illustrative example from the national context: India

The Indian Constitution lacks an explicit provision on the human right to water. However, the Constitution of India includes the right to life under its Article 21: 'No person shall be deprived of his life or personal liberty except according to procedure established by law.'[351] The Supreme Court and the lower Indian courts have on different occasions interpreted the right to life as safeguarding[352] the right to a healthy environment as they have explained that the right to life includes 'the right of enjoyment of pollution-free water and air for full enjoyment of life'.[353] Therefore the Courts have read the human right to water into the constitutional right to life. As with other national interpretations of the human right to water, the Indian formulation reflects the internal issues that India has regarding access to water. Pollution is one of the major threats to access to water and therefore the right to water is connected to a pollution-free environment.[354] In addition, as India is struggling to provide all of its citizens with access to potable water, protecting traditional water sources through the

345 Ibid. p. 189.
346 Ibid., p. 190.
347 Ibid.
348 (Committee on Economic, Social and Cultural Rights, 2002)
349 Inga Winkler, *The Human Right to Water Significance, Legal Status and Implications for Water Allocation* (First, Hart Publishing 2012). p. 55.
350 Jimena Murillo Chávarro, 'The Right to Water in the Case-Law of the Inter-American Court of Human Rights' (2014) 7 39. p. 47.
351 Www.cgsird.gov.in/constitution.pdf.
352 Subhash Kumar v. State of Bihar.
353 Subhash Kumar judgment at 424.
354 Inga Winkler, 'Judicial Enforcement of the Human Right to Water – Case Law from South Africa, Argentina and India' (2008) 1 Law, Social Justice & Global Development Journal 2 p. 3.

right to a healthy environment gives individuals the opportunity to facilitate their own access to water.[355]

The right to be free from cruel, inhuman or degrading treatment (Article 7 ICCPR)

The human right to water can also be derived from Article 7 ICCPR[356] which states: 'No one shall be subjected to torture or to cruel, inhuman or degrading treatment or punishment. In particular, no one shall be subjected without his free consent to medical or scientific experimentation.'[357] The Human Rights Committee explains in its communication that depriving individuals of food and water qualifies as cruel and inhuman treatment within the meaning of Article 7.

> *'The author has claimed, and the State party has not refuted, that he was deprived of food and water for several days after his arrest on 16 August 1988, tortured during two days after his transfer to the prison of Bata, and left without medical assistance for several weeks thereafter. The author has given a detailed account of the treatment he was subjected to and submitted copies of medical reports that support his conclusion. On the basis of this information, the Committee concludes that he was subjected to torture at the prison of Bata, in violation of Article 7; it further observes that the deprivation of food and water after 16 August 1988, as well as the denial of medical attention after the ill-treatment in the, or outside the, prison of Bata, amounts to cruel and inhuman treatment within the meaning of Article 7, as well as to a violation of Article 10, paragraph 1.'*[358]

This conclusion was arrived at due to a combination of factors as the author of the complaint was also detained and deprived of medical assistance and food.[359] Deriving the human right to water from Article 7 ICCPR could be restricted to those imprisoned or detained.[360] It is challenging to reach general conclusions concerning the characteristics of the human right to water under this article based on this Communication. However, it does demonstrate that the article provides protection for the human right to water, even if it is not a self-standing right.

An illustrative example from the national context: Botswana

In the national context, the right to be free from cruel, inhuman or degrading treatment has been interpreted to not only include the right to water for those being imprisoned

---

355 (Winkler, 2012) p. 111.

356 'The lack of access to sanitation can also be tantamount to inhuman or degrading treatment in certain circumstances, especially in the context of detention.' Report A/HRC/12/24 p. 15.

357 Article 7 ICCPR.

358 UN Human Rights Committee, *Essono Mika Miha v. Equatorial Guinea*, Communication No. 414/1990, 8 July 1994, para. 6.4 http://hrlibrary.umn.edu/undocs/html/vws414.htm.

359 UN Human Rights Committee, *Essono Mika Miha v. Equatorial Guinea*, Communication No. 414/1990, 8 July 1994, para. 6.4 http://hrlibrary.umn.edu/undocs/html/vws414.htm.

360 Murillo Chávarro. p. 48.

or detained. The Botswana Court of Appeal has held that a right to water can be derived from Section 7 of the Constitution.[361] This Section states that 'no person shall be subjected to torture or to inhuman or degrading punishment or other treatment.' The case concerned the San Nation residing in the Central Kalahari Game Reserve who had been refused permission by the government to recommission an existing borehole in order to safeguard their access to water. The Court of Appeal came to the conclusion that the government's refusal violated the San Nation's right not to be subjected to inhuman and degrading treatment.[362]

The above shows that the human right to water can be derived from the provisions of the ICCPR. This has many benefits, primarily the wider support that this instrument enjoys as opposed to the ICESCR. One of the disadvantages of the ICCPR, on the other hand, is that it does not allow for a self-standing right to water. Thus far, it has not been decided that an infringement of the human right to water in itself has taken place. Infringements have only been found in combination with violations of other rights.

The right of minorities to enjoy their own culture (article 27 ICCPR)

Article 27 says that ethnic, religious or linguistic minorities should not be denied the right to enjoy their own culture. Under this provision access to water can also be claimed however a few hurdles should be overcome.[363] The article can only be applied when individuals claim that water is necessary for cultural activities or other activities related to their religion. It should constitute a substantial compromise of the right to culture.[364] This makes claiming access to water under this right highly challenging. Also this right can also be called upon by minorities which makes it group-specific. However, as this right is part of the ICCPR it does have a legal status which could motivate enforcement.[365]

---

361 Bonolo Ramadi Dinokopila, 'The Right to Water in Botswana : A Review of the Matsipane' (2014) 11 African Human Rights Law Journal 282. p. 285.

362 Matsipane Mosethanyane & Others v The Attorney-General of Botswana High Court decision MAHLB-000393–09 (unreported). p. 63 https://www.escr-net.org/caselaw/2011/mosetlhanyane-and-others-v-attorney-general-botswana-civil-appeal-no-caclb-074–10.

363 See Daphina Misiedjan and Joyeeta Gupta, 'Indigenous Communities: Analyzing Their Right to Water under Different International Legal Regimes' (2014) 10 Utrecht Law Review 77.

364 See section on the right to take part in cultural live (article 15 ICESCR)

365 For an illustration of the application of this right for access to water purposes see: Human Rights Committee, Communication No. 1457/2006, Doc. CCPR/C/95/D/1457/2006, available on: http://docstore.ohchr.org/

The Government of Peru hindered the access to water for an indigenous community by firstly diverting the course of a river running past pastures and secondly by drawing groundwater significantly reducing access to water for the indigenous community. As a result, many animals died as a consequence of the gradual drying out of the wetlands. The complainant had exhausted all remedies at the national level and decided to petition the HRC arguing that the raising of llama's was a traditional custom of the Aymara people. And because the death of the animals was caused due to an infringement on their (customary) access to water, the government had infringed on their right to enjoy her own culture together with the other members of her group, in accordance with Article 27 of

3.3.2 LEGAL FOUNDATIONS OF THE HUMAN RIGHT TO WATER IN THE INTERNATIONAL COVENANT ON ECONOMIC, SOCIAL AND CULTURAL RIGHTS (ICESCR)

The International Covenant on Economic, Social and Cultural Rights was established in 1966, just like its twin, the ICCPR. As of May 2017, the ICESCR has 164 states parties. South Africa is the most recent state to have ratified the Covenant in April 2015. However, states such as the United States of America, Cuba and Myanmar have signed the Covenant but are not yet a party.[366] States which have only signed but not ratified a treaty are obliged to refrain from acts which defeat the object and purpose of that treaty.[367] However, ratification is still necessary for the treaty to become fully binding for that state and, depending on the national system of that state, become part of its domestic system.

A state which has ratified is obliged to: 'take steps, individually and through international assistance and cooperation especially economic and technical, to the maximum of its available resources, with a view to achieving progressively the full realisation of the rights recognised by the present Covenant by all appropriate means including particularly the adoption of legislative measures.'[368] In May 2013, the Optional Protocol to the ICESCR entered into force and it establishes inquiry and complaint mechanisms for the ICESCR.[369] It allows for the Committee overseeing the Covenant to consider complaints from individuals and groups who claim that a right under the Covenant has been violated. Thus far, the Protocol has 45 signatories and 22 states parties resulting in a very limited application of this procedure.[370] Nevertheless, this Protocol 'can surely contribute to improving the enjoyment of socioeconomic rights at the domestic level. In addition, the Optional Protocol will definitely recognise the right of victims of violations of social rights to be heard and give renewed hope to the millions of human beings who still do not enjoy the rights recognised in the Covenant.'[371]

The ICESCR does not include an explicit human right to water. Many argue that when this Covenant came about in 1966 water was not an imminent issue of concern

---

the Covenant. The HRC agreed stating that the Peruvian government has substantially compromised the water of life and culture of the complainant and their community.

366 See http://indicators.ohchr.org/

367 Articles 10 and 18 of the Vienna Convention on the Law of Treaties 1969.

368 Article 2(1) ICESCR.

369 See Catarina de Albuquerque, 'Chronicle of an Announced Birth: The Coming into Life of the Optional Protocol to the International Covenant on Economic, Social and Cultural Rights – The Missing Piece of the International Bill of Human Rights' (2010) 32 Human Rights Quarterly 144. on the history of the Protocol.

370 Https://treaties.un.org/Pages/ViewDetails.aspx?src=IND&mtdsg_no=IV-3-a&chapter=4&clang=_en.

371 Ibid. p. 177. In addition the 'international commitment to the human right to water and sanitation could be furthered by ratification of the Optional Protocol to the International Covenant on Economic, Social and Cultural Rights, which it has already signed.' Human Rights Council, 'Report of the Special Rapporteur on the Human Right to Safe Drinking Water and Sanitation, Catarina De Albuquerque Mission to Senegal (14 to 21 November 2011)'.

and therefore this right, other than the related right to food, was not included.[372] Nevertheless, this Covenant also includes several provisions from which the human right to water can be derived. Starting with Article 11 ICESCR, it states the following: 'the States Parties to the present Covenant recognise the right of everyone to an adequate standard of living for himself and his family, including adequate food, clothing and housing, and to the continuous improvement of living conditions. The States Parties will take appropriate steps to ensure the realisation of this right, recognising to this effect the essential importance of international co-operation based on free consent.'[373]

This provision sums up a list of the necessities for a life to be lived in dignity which includes several categories. However, the list is not exhaustive and leaves room for adding other basic needs that are of vital importance.[374] Water is undeniably of vital importance as without water a person will die of dehydration. The human right to water falls within Article 11 as both an independent, self-standing human right and as a condition for the fulfilment of other rights under this article. Article 11 points out the different aspects of the right to water. It covers water in its nourishment function for cooking and drinking purposes, fresh water of sufficient quality to sustain health and water for hygienic and sanitary needs. This section will discuss these aspects.

General Comment No. 15

In 2002, the Committee on Economic, Social and Cultural Rights presented its General Comment No. 15 which is not legally binding. However, this document 'extrapolates existing obligations from the nature of the Covenant.'[375] The CESCR argued that the right to water falls within the guaranteed categories which are essential to an adequate standard of living under Article 11.[376] 'The right to water is also inextricably related to the right to the highest attainable standard of health (Art. 12, para. 1) and the rights to adequate housing and adequate food (Art. 11, para. 1).'[377] The human right to water is not based on a specific separate article within the Covenant which makes defining the content somewhat challenging.[378] The Committee has however 'read in[379]' this right and

372 See Peter H Gleick, 'WATER IN CRISIS: PATHS TO SUSTAINABLE WATER USE' (1998) 8 Ecological Applications 571.

373 (Committee on Economic, Social and Cultural Rights, 2002)ICESCR article 11.

374 Eibe Riedel, 'The Human Right to Water and General Comment No. 15 of the CESCR' in Eibe Riedel and Peter Rothen (eds), *The Human Right to Water* (Berlines Wissenschafts – Verslag 2006). p. 25.

375 N Chowdhury and others, 'The Human Right to Water and the Responsibilities of Businesses: An Analysis of Legal Issues.' (2011) 3. p. 4.

376 (Committee on Economic, Social and Cultural Rights, 2002) par. 3.

377 (Committee on Economic, Social and Cultural Rights, 2002) par. 3.

378 The right to water is included in separate articles: Art. 14(h)(2) Convention on the Elimination of all Forms of Discrimination Against Women and Article 24(2) Convention on the Rights of the Child.

379 'Reading in' is an act of interpretation whereby a judicial or quasi-judicial body, following the object and purpose of a legal instrument, includes in its reading of a legislative instrument words or phrases that are not explicitly stated in the document. The effect is to read the instrument 'as if certain words appeared in the statute. The words are notionally 'included' to reflect in express, and therefore more

does not create any new rights by formulating this Comment.[380] General Comment No. 15 is currently the most authoritative exploration of the content of the right to water. The Committee has found a way to delineate an independent, self-standing human right to water by deriving it from the right to health and the right to an adequate standard of living.

General Comment No. 15 includes the following core statement: 'the human right to water entitles everyone to sufficient, safe, acceptable, physically accessible and affordable water for personal and domestic uses. An adequate amount of safe water is necessary to prevent death from dehydration, to reduce the risk of water-related disease and to provide for consumption, cooking, personal and domestic hygienic requirements.'[381] General Comment No. 15 explains that the right to water includes freedoms and entitlements. It includes elements which apply at all times, in all circumstances.[382] This document will often be mentioned while discussing the content of this right in the following sections.

The ICESCR provides several options. Most of these would result in the human right to water being considered as a subordinate right. Under those circumstances, it is up to the individual to prove that water is such an essential part of the right to food that, without it, this right cannot be enjoyed. The following sections discuss these options.

The right to food (Article 11 (1) and (2) ICESCR)

Article 11(1) stipulates that everyone has the right to adequate food as a part of the right to an adequate standard of living, while Article 11(2) establishes the right to be free from hunger.[383] In relation to the right to water, the right to food could provide a possible legal basis for a human right to water in three ways. Firstly, the right to water for drinking purposes as 'liquid food'; secondly, the right to water for the purpose of preparing food; and, lastly, the right to water for food production.[384] The former Special Rapporteur argued that 'the right to food includes not only the right to solid food, but also the right to liquid nourishment and to drinking water.'[385]

Human rights set minimum standards for individuals to live a life in dignity. Even though water is an essential component for producing food, water for food production and water to support health and sustain life are two separate issues in international human rights law. This is because, firstly, in order to fulfil the right to food, providing water to individuals locally is not required.[386] Secondly, the amount of water needed

---

readily observable form, the true construction of the words actually used, by way of a strained construction.' From James Spigelman, Statutory Interpretation and Human Rights 2008, p. 133–134.

380 Salman M. A. Salman; Siobhán McInerney-Lankford. p. 5.

381 (Committee on Economic, Social and Cultural Rights, 2002) par. 2.

382 (Committee on Economic, Social and Cultural Rights, 2003) par. 12.

383 A right to food encompasses the two norms of Article 11(1) and (2)taken together.

384 Kiefer and Brölmann. p. 196.

385 The right to food, Report by the Special Rapporteur on the right to food. 2001 E/CN.4/2001/53 para. 32.

386 Peter H Gleick, 'The Human Right to Water' (1998) 1 Water Policy 487. p. 491.

in order to provide every individual with enough water to grow their own food is substantially higher than ensuring water for basic cooking and hygienic purposes. Besides the quantity, another challenging factor would be the varying degrees of purity needed for agricultural purposes.[387] For this reason the right to water could be derived as an implicit right which serves as an condition for the right to food as mentioned in Article 11 ICESCR and Article 25 UHDR. This would mean that it is a subordinate right which does not enjoy an independent status. Also, the right to water derived from the right to food has a limited scope as it only includes water for drinking and cooking, while other purposes, such as water for personal hygiene, fall outside of its scope.[388] Interestingly, the issue of water in relation to the right to food was not mentioned in the previous General Comment No. 12 on the right to adequate food. Nevertheless, some argue that the human right to water should indeed include access to water for individuals' small scale agricultural activities.[389]

The right to housing (Article 11 ICESCR)

The right to an adequate standard of living in Article 11 ICESCR includes the right to housing.[390] The human right to water forms a part of the right to adequate housing as General Comment No. 4 explains:

> *'An adequate house must contain certain facilities essential for health, security, comfort and nutrition. All beneficiaries of the right to adequate housing should have sustainable access to natural and common resources, safe drinking water, energy for cooking, heating and lighting, sanitation and washing facilities, means of food storage, refuse disposal, site drainage and emergency services.'*[391]

This does not mean, however, that every household is required to have running water in order to fulfil the human right to water.[392] General Comment No. 15 explains that water services need to be physically accessible, meaning within a reasonable distance from the dwelling.[393] Therefore access to water under this article would also be considered a subordinate right. In addition, the right to water, as derived from the right to housing, would only cover those uses which can be related to housing such as domestic and personal uses.[394]

---

387 Riedel. p. 20.
388 (Winkler, 2015) p. 46.
389 This topic will be further discussed when discussing the content of the human right to water later in this chapter.
390 Article 14 CEDAW and Article 27 CRC also include a right to water.
391 General Comment No. 4: The Right to Adequate Housing (Art. 11) 1999 1. para. 8(b).
392 Catarina De Albuquerque and Virginia Roaf, 'Introduction' in Virginia Roaf, Inga Winkler and Muriel Schiessl (eds), *Realising the human rights to water and sanitation: A handbook by the UN Special Rapporteur Catarina De Albuquerque* (2014). p. 34.
393 (Committee on Economic, Social and Cultural Rights, 2002) par. 37.
394 (Winkler, 2012) p. 47.

An illustrative example from the national context: Argentina

The residents of informal settlements in central Buenos Aires were supplied with water by means of water tanker trucks and cisterns. This supply was disrupted in June, 2006 by the City of Buenos Aires. The residents applied for an injunction against the City at the local Administrative and Fiscal Court. In appeal, the Chamber for Administrative Matters, in relation to the right to housing, explained that 'the right to adequate housing includes the need for public services like water supply and basic sanitation.'[395] Also, the Court encouraged the City to improve its infrastructure: 'all water infrastructure should be of good quality and culturally adequate, taking into account gender, age and privacy needs of the people who live in the area.'[396]

The right to sanitation (Article 11 ICESCR)

Like the human right to water, the human right to sanitation is not explicitly recognised in the ICESCR.[397] The two rights have unsurprisingly been linked and tied together. Access to sanitation is crucial for avoiding disease-causing elements and is also directly tied to wastewater management.[398] These rights, however, are currently considered to be two separate and self-standing human rights.[399] The CESCR has found that:

> *'since sanitation is fundamental for human survival and for leading a life in dignity, the right to sanitation is an essential component of the right to an adequate standard of living, enshrined in Article 11 of the International Covenant on Economic, Social and Cultural Rights. The right to sanitation is also integrally related, among other Covenant rights, to the right to health, as laid down in Article 12 paragraphs 1 and 2 (a), (b) and (c), the right to housing, in Article 11, as well as the right to water, which the Committee recognized in its General Comment No. 15.'*[400]

The delinking of the right to water and the right to sanitation is motivated by the different characteristics and requirements of these two rights. Firstly, as sanitation is something of a taboo subject, it is often overshadowed by the realisation of the right to water.[401] Therefore delinking these subjects allows states to devote appropriate attention

395 Waterlex; and WASH United. p. 72.

396 Ibid. p. 72.
www.righttowater.info/rights-in-practice/legal-approach-case-studies/court-cases/ See: Argentina.

397 The right to sanitation is explicitly mentioned in Article 14.2h CEDAW and CRC Articles 24.1 and 24.2c.

398 'It is estimated that in Delhi, less than 20 percent of wastewater is treated before being dumped into the Yamuna River, while in Latin America, less than 14 percent of human waste is processed.' (Winkler, 2016) p. 1337.

399 See UNGA Resolution 70/169 which was adopted by consensus states that all individuals are entitled 'to have physical and affordable access to sanitation, in all spheres of life, that is safe, hygienic, secure, and socially and culturally acceptable and that provides privacy and ensures dignity.'

400 A/C.3/70/L.55/Rev.1.

401 Catarina De Albuquerque and Virginia Roaf, *On The Right Track. Good Practices in Realising the Rights to Water and Sanitation* (2012th edn, 2012) <www.ohchr.org/Documents/Issues/Water/BookonGoodPractices_en.pdf>. p. 213.

to sanitation by aiming to achieve subject-specific targets.[402] Secondly, water is not an essential element of the right to sanitation. Sanitation can be organised in such a manner that it does not require water.[403] Connected to this is the idea that water is more crucial than sanitation. Those who do not have access to water cannot find substitutes while those who do not have access to sanitation can still resort to open defecation.[404] Water is only required for wet sanitation services.[405] And the quality of the water used for these systems does not have to be fit for human consumption.[406] It can be concluded that even though they share the same legal foundations, the human right to water and sanitation are two separate rights and that water is not an essential element of the right to sanitation.

The right to health (Article 12 ICESCR)

The relationship between (the right to) water and right to health is a strong one. The absence of water has detrimental effects on human health and can lead to death. Diseases and ailments such as diarrhoea and cholera are caused by lack of access to safe drinking water. According to the WHO diarrheal diseases are the 7th leading cause of death in the world.[407] In addition to guaranteeing access to safe water, access to sanitation is of great importance for the realisation of the right to health. Article 12 ICESCR is the most comprehensive provision on the right to health in international human rights law.[408] This provision not only creates obligations concerning access to drinking water but also creates obligations which apply to water management as the next sections will explain.

Article 12 states that:

1. The States Parties to the present Covenant recognize the right of everyone to the enjoyment of the highest attainable standard of physical and mental health.
2. The steps to be taken by the States Parties to the present Covenant to achieve the full realization of this right shall include those necessary for:
   (a) The provision for the reduction of the stillbirth-rate and of infant mortality and for the healthy development of the child;

---

402 (Ellis & Feris, 2014) p. 617; Pedi Obani and Joyeeta Gupta, 'Human Right to Sanitation in the Legal and Non-Legal Literature: The Need for Greater Synergy' (2016) 3 Wiley Interdisciplinary Reviews: Water <http://doi.wiley.com/10.1002/wat2.1162>. p. 683 – 684.

403 (De Albuquerque & Roaf, 2012) p. 149.

404 Those who resort to open defecation are often marginalized. See example: Kathleen O'Reilly, Richa Dhanju and Abhineety Goel, 'Exploring "The Remote" and "The Rural": Open Defecation and Latrine Use in Uttarakhand, India' (2017) 93 World Development 193.

405 What could be argued here is that once individuals have access to wet sanitation services on the basis of non-retrogression and culture appropriateness, these systems should be maintained.

406 (Obani & Gupta, 2016) p. 684.

407 Www.who.int/mediacentre/factsheets/fs310/en/
See also William Onzivu, 'The Long Road to Integrating Public Health into Sustainable Development of Shared Freshwaters in International Environmental Law: Lessons from Lake Victoria in East Africa' (2012) 46 International Lawyer 867. p. 871.

408 General Comment No. 14 The right to the highest attainable standard of health 2000 1. paragraph 2.

(b) The improvement of all aspects of **environmental and industrial hygiene**;
(c) The prevention, treatment and control of epidemic, endemic, occupational and other diseases;
(d) The creation of conditions which would assure to all medical service and medical attention in the event of sickness.

General Comment No. 14 on the right to health underlines the importance of water to the right to health. It sees access to water as a determinant of health. The Committee has adopted a broad approach when interpreting the right to health. It states:

> *The Committee interprets the right to health, as defined in Article 12.1, as an inclusive right extending not only to timely and appropriate health care but also to the underlying determinants of health, such as access to safe and potable water and adequate sanitation, an adequate supply of safe food, nutrition and housing, healthy occupational and environmental conditions, and access to health-related education and information, including on sexual and reproductive health.*[409]

Interpreting the right to water as being a part of the right to health contributes several aspects to the human right to water. Firstly, it implies that health standards must be set for potable water.[410] Secondly, not only the water itself has to be safe and not harmful to a person's health but also the general environment of that person must be clean and safe.[411] And thirdly, the link between the right to water and health allows for the issue of sanitation and personal hygiene to be kept in focus.[412]

An illustrative example from the national context: Bangladesh

Supported by international agencies, Bangladesh encouraged its citizens to move from using surface water, which was considered to be contaminated, to using groundwater.[413] As a result many tube wells were installed. However the groundwater had not been tested for arsenic contamination which can be naturally occur. When many fell ill due to arsenic poisoning and it became clear that this was due to the consumption of groundwater, a former member of parliament filed a public interest petition against the Government and other responsible public authorities. The case eventually ended up at the Court of Appeal. Here the Court did not exclusively base its

409 Ibid. paragraph 11.
410 Paul Hunt and Gunilla Backman, 'Health Systems and the Right to the Highest Attainable Standard of Health', *Human Rights and Climate Change* (Cambridge University Press 2009). p. 247.
411 Elliot Curry, 'Water Scarcity and the Recognition of the Human Right to Safe Freshwater' (2010) 9 Northwestern Journal of International Human Rights 103. p. 118.
412 Riedel. p. 27.
413 See Farhana Sultana, 'Water, Water Everywhere, But Not a Drop to Drink: Pani Politics (Water Politics) in Rural Bangladesh' (2007) 9 International Feminist Journal of Politics 494 <www.tandfonline.com/doi/abs/10.1080/14616740701607994>.

ruling on the right to health but did explain that the right to health include the right to drinking water.[414]

The right to a healthy natural environment (Article 12(2)b)

Neither the ICCPR nor the ICESCR mention the right to a healthy environment. Nevertheless, it has been claimed that this right falls within the scope of the right to life, the right to health and the right to privacy[415] mentioned in global and regional instruments.[416] The Committee on Economic, Social and Cultural Rights explains in General Comment No. 14 that the right to health includes the right to a healthy environment. Here it elaborates on the provision in Article 12.2(b) by mentioning several measures which should be taken in order to improve 'all aspects of environmental and industrial hygiene'.[417] Among them is preventing and limiting the exposure of the population to harmful substances and 'other detrimental environmental conditions that directly or indirectly impact upon human health'.[418]

In General Comment No. 15 on the right to water the Committee takes a similar approach and applies the concept of environmental hygiene to safe water resources. It states that States are obligated to prevent unsafe and toxic water conditions. 'For example, States parties should ensure that natural water resources are protected from contamination by harmful substances and pathogenic microbes.'[419] The Committee does not elaborate on a state's obligations regarding water resources and management beyond this comment. This can be explained by the Committee focusing merely on paragraph 1 of Article 12 ICESCR. It states that the right to water emanates from a number of rights including the right to an adequate standard of living and is 'also inextricably related to the right to the highest attainable standard of health (Art. 12, para. 1).'[420] Nevertheless, the Committee does refer to its explanation on the right to a healthy natural environment in paragraph 15 of General Comment No. 14.

An illustrative example from the national context: Colombia

The inhabitants of Los Llanos de Cuiva in Colombia enjoyed access to water from a community-managed aqueduct. The water from this aqueduct became unsafe after a pig

414 Rabia Bhuiyan v. Ministry of LGRD and others, Appellate Division (Civil Appeal no 118 of 1999), 27 August 2005. Paragraph 18 Available at: www.globalhealthrights.org/wp-content/uploads/2013/02/SC-2007-Rabia-Bhuiyan-v.-LGRD.pdf.

415 Discussing the right to privacy falls outside the scope of this research. On claims concerning the right to a healthy environment under the right to privacy see.

416 Sueli Giorgetta, 'The Right to a Healthy Environment, Human Rights and Sustainable Development' (2002) 2 International Environmental Agreements: Politics, Law and Economics 173 <http://link.springer.com/article/10.1023/A:1020938009559>. p. 177.

417 General Comment No. 14 The right to the highest attainable standard of health. par. 15.

418 Ibid. par. 15.

419 (Committee on Economic, Social and Cultural Rights, 2002), paragraph 8.

420 (Committee on Economic, Social and Cultural Rights, 2003), paragraph 3.

farmer settled near the main water basin and the runoff from the farm contaminated the water with faecal matter. Communities who made use of the water suffered different negative health consequences. Children, for instance, suffered skin and gastrointestinal infections. The Municipal Civil Court decided that the right to a healthy environment includes the protection of water resources as a necessary part of the national policy on renewable natural resource. This policy includes the sustainable use of natural resources for the benefit of present and future generations.[421] Because this all facilitates the ability to consume clean water. The Court also decided that water services which are managed by the community itself still fall under the obligation to protect from infringements by third parties.

The right to take part in cultural life (Article 15 ICESCR)

In many religions and cultures, water plays a central role due to its purifying and life-sustaining qualities.[422] These qualities assist in considering water to be sacred.[423] Especially for indigenous people, water bodies are sacred sites which form a significant part of the cultural experience of the indigenous population.[424] Because of its significant role in the cultural life of individuals it is not far-fetched to claim that, for certain individuals, access to water is an important part in order to partake in cultural life. However, this type of human right to water would be classified as a condition or a subordinate right and not as an independent right. Kiefer has reasoned that the right to water stemming from the right to take part in cultural life is not an independent right because '[c]onsidering the diversity of occupational and cultural practices around the world, the realisation of these rights simply does not *per se* require access to water.[425]' Individuals would have to prove that water is an essential part of their culture and that their culture cannot be enjoyed without it and that no substitutes are possible. The right to partake in cultural life has to be fully undermined. Another consequence of a right to water for cultural practices would for instance be that it would be subordinate to the right to use water for domestic and personal uses.[426]

An illustrative example from the national context: Chile

There are several cases which indirectly point at a human right to water being delineated from cultural rights. For instance in *Alejandro Papic Domínguez con Comunidad*

421 Waterlex; and WASH United. p. 122.

422 Nan Palmer, 'Water Adequacy and Essential Element for Empowering People' (2012) 19 ILSA Journal of International and Comparative Law 31. p. 37.

423 Other cultural and religious groups attribute a central role to water. For instance, religious and spiritual practices illustrate the important and sacred role that water plays. For instance in Christianity water is used during baptisms, while symbolising purification and the rejection of original sins. Buddhists use water during funerals and Muslims use water to purify the body before prayer and touching the Koran. Nan Palmer, 'Water Adequacy as an Essential Element for Empowering People' (2012) 19 ILSA Journal of International and Comparative Law 31. p. 38.

424 (Tauli Corpuz, 2006) p. 29.

425 Kiefer and Brölmann. p. 198.

426 ESCR Committee, General Comment No. 15, U.N. Doc. E/C.12/2002/11, paragraph 6.

Indígena Aimara Chusmiza y Usmagama[427], where customary water right were recognised in favour of indigenous peoples in a situation where water rights were also registered to a third party. The Court of Letters (first instance) held that Constitution does not only protect water rights as registered under the Water Code but also water rights for customary uses. The Supreme Court upheld the judgment however, as the government has failed to enforce the decision by the Supreme Court, the case has been brought before the Inter-American Commission on Human Rights. In the petition to the Commission the Aimara de Chusmiza-Usmagama community expressed that access to water is among other things fundamental to among other things the development of their culture and their spiritual life.[428]

The right to development

Water is highly relevant for the realisation of many human rights, as mentioned above. Also, without water, development cannot be pursued. For this reason, considering the right to water as a component of the right to development is not surprising. The Declaration on the Right to Development was adopted in 1986 by way of a resolution which makes its provisions non-binding as far as states are concerned.[429] Article 1 of the Declaration reads: 'The right to development is an inalienable human right by virtue of which every human person and all peoples are entitled to participate in, contribute to, and enjoy economic, social, cultural and political development, in which all human rights and fundamental freedoms can be fully realized.'

The right to development connects human rights to development. It emphasises development as a human-centred participatory process. The UN Declaration on the Right to Development provides a foundation for the improvement of excluded groups such as indigenous peoples.[430] The right to development, as it also includes a right to a process of development, combats the structural reasons for poverty and the lack of control over resources. As ways to attain this improvement, meaningful participation, non-discrimination and the distribution of the benefits of development are considered to be relevant.[431] The characteristics of the right to development seem to be unclear, however.

427 Alejandro Papic Domínguez con Comunidad Indígena Aimara Chusmiza y Usmagama [2009] Corte Suprema 2.840–2008 available at: www.politicaspublicas.net/panel/jp/439–2009-chusmiza.html.

428 Inter-American Commission on Human Rights, Report N° 29/13, Petition 1288–06, March 20, 2013.

429 Resolution 41/218 www.un.org/documents/ga/res/41/a41r128.htm.

430 The right of indigenous peoples within the right to development can be linked to several international legal standards and principles and reaffirms other rights, such as participation rights, the right to self-determination and rights concerning land and natural resources. The right to development is also connected to the enjoyment of the right to water: One of the most important causes of poverty within indigenous communities is the denial of access and control over their own resources.

431 Salomon (2003) p. 3.

The right to development is often seen as a right for nation states.[432] Nonetheless, both individuals as well as a people (groups) can invoke this right.[433] In this regard, Filmer-Wilson argues that states are obliged to provide water to both individuals as well as communities and clearly identifiable groups such as women.[434] This approach would allow for future generations to have their access to water covered by the right to development. In order to give that right more teeth, a further interpretation of and an agreement on its scope and legal content are needed.[435] The concept of a right to development is still, at the moment, 'soft law'.

An illustrative example from the national context: Costa Rica

The Municipality of Poás permitted earthworks to be executed, in an area where aquifers are highly susceptible to contamination. The municipality negated obtaining an environmental impact assessment by the Ministry of Environment and Energy. However this was later granted. The applicant argued that the administrative authorisation for the construction showed inconsistencies and violated several provision of the Political Constitution of Costa Rica. Also in this case, did the Supreme Court of Justice not exclusively base its conclusion that the right to water was infringed upon on the right to development. However it asserted that access to drinking water is associated with the development and socio-economic growth of the population.[436]

Conclusions on the legal foundations of the human right to water

From the previous examination it can be concluded that the human right to water can be derived from several provisions in human rights treaties. These different provisions give different characteristics to the human right to water. What should be kept in mind is that these different characteristics can assist individuals under different circumstances. Deriving the right from Article 6 ICCPR provides the right with a wider application and a strong enforcement mechanism. However, the scope of the right is very limited. For communities who are in life-threatening situations, where they, for instance, have been relocated to a place where water is inaccessible, this article may be the most relevant.

---

432 Margot Salomon and Arjun Sengupta, 'The Right to Development: Obligations of States and the Rights of Minorities and Indigenous Peoples' (2003). p. 4.

433 Felix Kirchmeier, 'The Right to Development – Where Do We Stand ? State of the Debate on the Right to Development Dialogue on Globalization', vol 23 (2006). p. 10.

434 Emilie Filmer-Wilson, 'The Human Rights-Based Approach to Development: The Right to Water' (2005) 23 Netherlands Quarterly of Human Rights 213.

435 Kirchmeier. p. 13.

436 Corte Suprema de Justicia (2004): Comité Pro-No Construc- ción de la Urbanización Linda Vista, San Juan Sur de Poás c/ Ministerio de Ambiente y Energía y Otros Cons. VI.

3.3.3 LEGAL FOUNDATIONS OF THE HUMAN RIGHT TO WATER WITHIN CUSTOMARY LAW

In order for the human right to water to be part of customary law, two criteria should be taken into account according to Article 38(1)(b) of the Statute of the International Court of Justice, namely: there needs to be a general practice and states need the conviction that this practice is accepted as law. This section analyses the current state of the human right to water as customary law. It does so by discussing the practice of states and the beliefs or convictions that are connected to this state practice. In order to deduce state practice this section studies states' behaviour when it comes to the human right to water. This behaviour is demonstrated by both action and inaction. The opinio juris is studied according to the public statements made by states, such as declarations, resolutions, etc., which reflect the states' belief that they are legally bound.[437] It should be kept in mind that rights within customary international law develop gradually. No one action creates a custom.[438] Therefore multiple forms of evidence are necessary to prove that a certain right is part of customary law.

Some argue that in evaluating the status of a right under customary law, one of the elements should weigh heavier than the other. For instance, some argue that the *Nicaragua* case represents the view that statements weigh heavier than practice. In this case the International Court of Justice referred to a relevant resolution using it as evidence for the ban on the use of force and the principle of non-interference.[439] This approach will not be followed in this section as the position will be taken that both elements for classifying customary law should be equally weighed. The current opinion in the literature is that the right to water has not yet developed as part of customary law.[440] This section will conclude the same. It will argue that the right to water is still materialising in customary law as it currently has a weak status, which is mostly fuelled by state practice in combination with states' statements.[441]

3.3.3.1 State practice

It has become clear that the number of states implementing the human rights to water at the domestic level and supporting the right at international level has drastically

---

437 Thielborger. p. 234.

438 The ICJ has rejected the idea of an instant custom in several instances. See North Sea Continental Shelf (Federal Republic of Germany v Denmark; Federal Republic of Germany v The Netherlands) Judgment, ICJ Reports 1969, 3 at para 73; Military and Paramilitary Activities in and Against Nicaragua (Nicaragua v United States of America) Merits, Judgment, ICJ Reports 1986, 14 at para 184.

439 As the Court stated in the case concerning Military and Paramilitary Activities in and against Nicaragua (Nicaragua v. United States of America) (I.C.J. Reports 1986, p. 94, para. 176): "there is a specific rule whereby self-defence would warrant only measures which are proportional to the armed attack and necessary to respond to it, a rule well established in customary international law".

440 See Bates. and Cavallo.

441 Pierre Thielbörger, 'The Right to Water: Effective Multi-Level Protection of a Multi-Faceted Human Right? – An Application of the Kadi and Medellin Approaches to the Case of the Right to Water'. p. 95.

increased.[442] But in order for these actions to be classified as state practice, the practice must be uniform, consistent, general and of a certain duration.[443] This is a very high threshold as it means that states must behave almost identically when it concerns the human right to water. Here one of the challenges is the question of what can be considered as behaviour: what the state actually does or what the state says it does. When we look at the actual behaviour of states it can be concluded, by looking at the state of access to water for individuals globally, that there is still room for improvement by states.[444] Actions such as disconnecting those who are not able to afford payment and the unregulated pollution of water resources illustrate that there is, as yet, no consistent state practice when looking exclusively at the behaviour of states.[445] However, it can be argued that what is also relevant is whether states have expressed their commitment to the right in resolutions, policy statements etc.[446]

Actions which are contrary to these statements should then be regarded as human rights violations and not as behaviour influencing state practice.[447] The solution to reconciling these two perspectives is to look at whether states consider their actions to be legally justified or whether they consider them to be human rights violations.[448] Evidence for this can be found in their behaviour and statements as well. For instance, if a state is concealing its actions or denying them, this is a clear gesture that it understands its actions as violations. If, on the other hand, a state publically recognises its actions and finds justifications for them, it can be considered that the state does not see the realisation of the right as part of state practice.

In order to decide whether or not the human right to water is part of customary law, this section will balance the two perspectives, reconciling both state behaviour as well as its statements. It does so as only depending on one of these elements will leave a distorted view. If only the statements made by states were considered to establish state practice, states could be immediately legally bound without having the time to develop their policies. On the other hand, if only states' behaviour would bring about customary law, states would not be held responsible for the statements they make as long as their practice is still contrary to their statements. Therefore the following section will discuss whether access to water is consistently approached as a human rights issue by analysing

442 See Waterlex; and WASH United; Thorsten Kiefer and others, *The Human Right to Safe Drinking Water and Sanitation in Law and Policy – a Sourcebook* (WASH United, Freshwater Action Network, Waterlex 2012).

443 See Bruno Simma and Philip Alston, 'The Sources of Human Rights Law: Custom, Jus Cogens, and General Principles' [1988] Australian Yearbook of Law 82.

444 Winkler, *The Human Right to Water Significance, Legal Status and Implications for Water Allocation.* p. 68.

445 Simma and Alston. p. 104.

446 Joyeeta Gupta, Rhodante Ahlers and Lawal Ahmed, 'The Human Right to Water : Moving Towards Consensus in a Fragmented World' (2010) 19 Review of European Community and International Environmental Law 294. p. 294.

447 (Winkler, 2012) p. 69.

448 Cavallo. p.

statements, resolutions, domestic legislation and other relevant documents on the right to water.[449]

The interaction between domestic law and international law is highly relevant in the field of human rights. The domestic level also contributes to the development of law at the international level in general and customary law as is relevant to this topic.[450] There are two important common factors between domestic human rights and international human rights. Firstly, national constitutions often include human rights which follow the language used in the UDHR and other treaties. Therefore for an interpretation of such rights we can look at the international level.[451] Secondly, human rights at both the domestic and international level are concerned with the same relationship, that of the state and the individual. For these reasons the influence of domestic legislation on the development of international law is relevant.[452]

States are increasingly and explicitly including the human right to water in their domestic legislation. Examples can be seen in constitutions and lower legislation. For instance, in its National Water Law[453] Paraguay has stated that access to water is a fundamental right. And in Article 26 of its Water Act Mozambique has given priority to water allocation for human consumption and sanitation.[454] States like India and South Africa are considered to be frontrunners in implementing the human right to water in domestic legislation which has also been backed up by the judiciary.[455] They illustrate two different approaches. South Africa has included a 'justiciable, affirmative right of access to adequate water, a right enshrined in the country's constitution'[456] which has been upheld by the country's Constitutional Court. The South African formulation has been successful in requiring the state to take reasonable legislative measures.[457] 'The key to the South African model is that, while the courts do not decide the rights that should be legally protected, they retain the power to review the "reasonableness" of policies implementing such protections.'[458]

India has taken a different approach and has chosen to interpret the right to life as formulated in the Constitution in such a way that it implicitly recognises a justiciable

449 Www.ohchr.org/Documents/Publications/training9chapter1en.pdf.

450 See Ralph P Hall, Barbara Van Koppen and Emily Van Houweling, 'The Human Right to Water: The Importance of Domestic and Productive Water Rights' (2013) 20 Science and engineering ethics <www.ncbi.nlm.nih.gov/pubmed/24337891> accessed 25 July 2014.

451 (Cavallo, 2012) p. 200.

452 Cavallo, 2012) p. 200.

453 Philippe Cullet and others, *Water Law for the Twenty-First Century National and International Aspects of Water Law Reform in India* (2010). p. 14.

454 Malgosia Fitzmaurice, 'The Human Right to Water' (2006) 18 Fordham Environmental Law Review 537. p.563.

455 Ibid. p. 573 and see Duncan French and Nathan John Cooper, 'The Right to Water in South Africa: Constitutional Managerialism and a Call for Pluralism' (2012) 5 <http://eprints.lincoln.ac.uk/6738/>.

456 Anonymous, 'What Price for the Priceless? Implementing the Justiciability of the Right to Water' (2007) 120 Harvard Law Review 1067. p. 1068.

457 See for instance Mazibuko and Others v City of Johannesburg and Others (Centre on Housing Rights and Evictions, Amicus Curiae) Constitutional Court of South Africa (CCT 39/09) [2009] ZACC 28; 2010 (3) BCLR 239 (CC); 2010 (4) SA 1 (CC) 8 October 2009.

458 (Anonymous, 2007) p. 1088.

right to water. At both state and federal level, courts have followed this approach. 'Though there seem to be many benefits to the Indian model of justiciability, including giving governments the flexibility of not explicitly obligating themselves to provide access to water, there are also drawbacks. First, deriving a justiciable right to water from the broad right to life greatly limits the extent of justiciability. Just as the right to life has traditionally been considered a negative right, placing upon the State the responsibility of not taking a citizen's life, so the right of access to water under the Indian model seems limited to protecting the citizen from interference with access to water.'[459]

The former Special Rapporteur on the human right to water, Catarina de Albuquerque, has amassed 'good practices' worldwide, illustrating that many states are working towards realising the human right to water for their citizens.[460] These good practices include Kenya's water reform. This state has included an explicit recognition of the right to water and sanitation in its Constitution and has developed several legislative instruments to expand services to those who were previously unserved.[461]

Even though these examples of the inclusion of the human right to water at the domestic level have shown an increase during the last decade, the amount of states undertaking these actions is still a minority. Around 40 countries, to date, have taken such steps towards national recognition while the UN has 193 member states.[462] Therefore it can be concluded that there are currently not enough states recognising the human right to water at the domestic level for the human right to water to be considered as state practice.

3.3.3.2 Opinions

States have made statements within different forums about their perception of water. Within these forums different viewpoints can be noticed around water,, ranging from seeing water as a mere economic commodity, to recognising it as a need and finally to recognising it as a human right.[463] These statements have not only been made within the field of human rights but also concerning international environmental law[464] and they contribute to the development of the human right to water as customary law as they reflect states' opinions and possible behavior.[465] They are often considered to be

---

459 Ibid. p. 1088.

460 De Albuquerque and Roaf, *On The Right Track. Good Practices in Realising the Rights to Water and Sanitation.*

461 (De Albuquerque & Roaf, 2012) p. 65.

462 Waterlex; and WASH United.

463 International Conference on Water and Environment, Dublin, Ir., Jan. 26–31, 1992, The Dublin Statement on Water and Sustainable Development, at 7 (1992). There are three features connected to the commodity approach: water should be priced, water delivery should be opened to the market and water resources.

464 See for instance the UN Convention on the Law of the Non-navigational Uses of International Watercourses, available at http://untreaty.un.org/ilc/texts/instruments/english/conventions/8_3_1997.pdf.

465 Opinions range from mere statements to statements which are made in response to actions.

'soft law' as they lack legal authority.[466] Their varying forms and levels of legal authority are nevertheless valuable contributions to the development of customary law.[467] However, as mentioned before no one statement makes customary law, there needs to be a multitude of similar opinions. The following section delves into the influential statements which have been made on international water and development policy.

The first statements on the right to water where a significant number of states were involved were at the 1972 UN Conference on the Human Environment in the Stockholm Declaration.[468] This conference took a human rights-based approach to environmental protection. Here it was declared that: 'The natural resources of the earth including the air, water, land, flora and fauna and especially representative samples of natural ecosystems must be safeguarded for the benefit of the present and future generations through careful planning or management, as appropriate.'[469]

The Stockholm Declaration was followed by the Mar del Plata Action Plan. In the Mar del Plata Declaration it is stated that 'all peoples whatever their stage of development and their social and economic conditions, have the right to have access to drinking water in quantities and of a quality equal to their basic needs.'[470] Even though it is not legally binding, this statement can be seen as the genesis of the human right to water. Nevertheless, the formulation of this statement leaves a few gaps which make distinguishing water as an human right challenging. Firstly, the statement uses the term 'peoples' which is mostly used to refer to a state as the representative of the peoples or refers to peoples in general, meaning the world population.[471] However, the statement also refers to 'basic needs' which assumes that the statement is focused on individuals. This statement nevertheless lacks an explicit reference to an individual right to water.

Mar del Plata was followed by other initiatives such as the Dublin Statement where water was recognised as an economic good, but not overlooking its role as a 'basic right of all human beings to have access to clean water and sanitation at an affordable

466 Jordan Daci, 'Protection of the Human Right to Water Under International Law The Need for a New Legal Framework' (2012) 6 Academicus International Scientific Journal 71. p.74.

467 John C Dernbach and Joel a Mintz, 'Environmental Laws and Sustainability: An Introduction' (2011) 3 Sustainability 531 <www.mdpi.com/2071-1050/3/3/531/> accessed 12 February 2014. p. 539.

468 Declaration of the United Nations Conference on Human Environment, available at: www.unep.org/Documents.Multilingual/Default.asp?documentid=97&articleid=1503 last visited on December 9th, 2012.

469 Principle 2 of the Stockholm Declaration.

470 Report of the United Nations Water Conference, Mar del Plata, 14–25 March 1977 (United Nations publication, Sales Np. E.77.II.A.12), part 1, chapter I, section C, para. 35. See Bourquain. p. 12.

471 'Peoples' are mostly used with reference to the right to self-determination. As a legal principle, the right of self-determination can be regarded as containing at least five elements: first, the right of a people of an existing state to determine freely their political status as well as pursue their economic, social and cultural development without external interference; second, the right of a people under foreign rule to free themselves; third, the right of a people to secede; fourth, the right of a people to determine their constitution including autonomous status, and finally, the right to govern by way of democracy. Roepstroff (2004), p. 17 Also see: Rosas, Allan, "Internal Self-Determination", in Tomuschat, Christian (ed.) Modern Law of Self-Dertermination (Dordrecht 1993) and Cassesse, Antionio, Self-determination of peoples: a legal reappraisal.

price.'[472] This statement which was included in Principle 4 was adopted not by states but was formulated by 500 experts designated by states as part of four guiding principles which are highly influential. This statement clearly recognises water as a human right.[473]

In 1992, Agenda 21 was adopted and it includes a chapter on drinking water supply and sanitation. In this chapter the Agenda refers to the Mar del Plata Action Plan and considers the right to have access to drinking water to be a 'commonly agreed premise'.[474] In 2002 the UN members adopted the Millennium Declaration, from which the Millennium Development Goals (MDGs) were derived. MDG 7 calls on states to ensure environmental sustainability and includes the target of halving the number of people who do not have sustainable access to safe drinking water by 2015.[475] This document has been an influential document for international policy.[476] However, the formulation supports access to water for individuals but does not explain access to water as an entitlement.[477]

One of the central UN documents which shines light on the opinions of states concerning the human right to water is the UN General Assembly resolution which was adopted in July 2010.[478] This resolution was the first in which the General Assembly and all UN states were able to express their viewpoint. Therein the UNGA states that it 'recognizes the right to safe and clean drinking water and sanitation as a human right that is essential for the full enjoyment of life and all human rights.'[479] The GA specifically chose the wording 'recognise' instead of, for instance, 'declares' which demonstrated that it found that a new right had not been created but that an already existing right was being recognised.[480]

This resolution received a great deal of support. The Bolivian government took up the reins of leadership and found 39 states to co-sponsor it. Amongst them were Bangladesh, the Dominican Republic and Yemen. Even though many other states expressed their support, the resolution could not be adopted by consensus.[481] Instead the United States called for a vote as it found that the drafting process had not been transparent and that the resolution could undermine the process that was ongoing at the Human Rights Council where a resolution concerning the same topic was at the negotiation stage. It stated the following: 'The United States regrets that this draft

---

472 Salman 2004, p. 9 Dublin Statement on Water and Sustainable Development, see Journal of Water SRT Aqua, Vol. 41, No. 3, at 129.

473 Gleick, 'The Human Right to Water'. p.493.

474 Chapter 18.47.

475 Target 7.c.

476 PB Anand, 'Right to Water and Access to Water: An Assessment' (2007) 19 Journal of International Development 511. p. 524.

477 Ibid. p. 518.

478 UN General Assembly Resolution on the human right to water and sanitation (2010), A/64/L.63/Rev.1, available on www.un.org/ga/search/view_doc.asp?symbol=A/RES/64/292.

479 Ibid par. 1.

480 (Winkler, 2012) p. 78.

481 General Assembly 2010. 108th Plenary Meeting: Agenda Item 48 (A/64/PV.108). United Nations (28 July).

resolution diverts us from the serious international efforts under way to promote greater coordination and cooperation on water and sanitation issues. The draft resolution attempts to take a short cut around the serious work of formulating, articulating and upholding universal rights. It was not drafted in a transparent, inclusive manner, and the legal implications of a declared right to water have not yet been carefully and fully considered in this body or in Geneva.'[482]

The eventual outcome of the vote on the resolution was as follows; 122 in favour, 41 abstentions, while 29 states were absent during the voting.[483] Even though the United States expressed concern about the resolution as it supported the process at the Council, a majority of states, which were also Human Rights Council members (36 of the 47 members), voted in favour of the GA resolution.[484] While only 7 states abstained, among them were Austria, the United States and Poland.[485] This result shows that in the opinion of states a conflict between the GA and the HRC on this topic is not an absolute fact.[486] Some states chose to abstain from voting but did not vote against the resolution as this would have been an unpopular decision.

There were many different reasons as to why states chose to abstain, besides reasons connected to the Human Rights Council process. For instance, some were concerned about the consequences that the resolution might have, while others found the drafting process to lack transparency.[487] Overall the adoption of this GA resolution is seen as a major step forward in recognising water as a human right and sheds some light on states' political will and statements. However, the resolution did fall short in explicitly establishing the right to water as an independent human right.[488] In addition, the resolution was also not adopted unanimously, as 70 states abstained meaning that the resolution had less support than it would have done if it had been adopted unanimously. Nevertheless, this resolution is of great importance for the development of the human right to water.

As mentioned, parallel to the process at the UNGA another process concerned with the human right to water was also underway. The UNHRC is the highest human rights organ in the UN and 'addresses violations, promotes human rights assistance and education, reviews States' human rights records, works to prevent human rights abuses, responds to emergencies, and serves as an international forum for human rights

---

482 UN General Assembly, 108th plenary meeting, agenda item 48, debate concerning the resolution on the human right to water and sanitation A/64/PV.108, p. 7 available on: www.un.org/ga/search/view_doc.asp?symbol=A/64/PV.108.

483 UN Press release 28 July 2010, General Assembly adopts resolution recognizing access to clean water, available on: www.un.org/News/Press/docs/2010/ga10967.doc.htm.

484 Ibid.

485 Ibid.

486 The behaviour of the states has shown that they do not see a conflict. A reason for the disagreement can be based on????

487 Ibid. p. 77.

488 Maria Gavouneli, 'A Human Right to Groundwater?' (2011) 13 International Community Law Review 305 <http://ezp-prod1.hul.harvard.edu/login?url=http://search.ebscohost.com/login.aspx?direct=true&db=aph&AN=63526103&site=ehost-live&scope=site&scope=cite> accessed 8 March 2013. p. 306.

dialogue.'[489] In September, 2010 the Council affirmed that the human right to water is a legally binding right. It found that the right is derived from the right to an adequate standard of living which, in turn, is derived from the ICESCR and other treaties such as the Convention of the Rights of the Child.[490]

The UNHRC resolution was put forward by Germany and Spain and had 50 co-sponsors. Amongst them were the Netherlands, Bosnia and Armenia. Interestingly, some of the co-sponsors of the UNHRC resolution like the Netherlands had abstained from voting on the UNGA resolution. A reason for this could have been the difference in procedures.[491] The UNHRC resolution therefore adds to the idea that the human right to water is more widely supported than the vote at the GA demonstrated.[492] Another aspect that this resolution adds it that it puts the human right to water within the sphere of legally binding human rights instruments.

Also, the SDGs adopted in 2015 by all 193 UN member states is another expression of the commitment to realising universal access to water, even though it is not legally binding. It creates another reason for states to recognise and implement the human right to water.[493]

From the above-mentioned analysis it can be concluded that the human right to water as part of customary law is in full development. As far as it concerns water needed to sustain life as is covered by Article 6, it could be said that the human right to water deriving from this article has become part of customary law. Having the human right to water forming part of customary law is important as it creates obligations for all states including those which are not parties to treaties which explicitly or implicitly recognise the human right to water.[494]

State practice as it currently stands is still too minor for it to be consistent and universal. However, if the trend of including the human right to water in domestic legislation continues, then state practice might steadily move forward. On the other hand, the opinion surrounding the human right to water does hint at there being universal acceptance of the human right to water. States may not have reached full consensus on the content and the obligations stemming from this right but the existence thereof has been confirmed by the majority of states.

---

489 The Human Rights Council and the role of the Office of the High Commissioner for Human Rights available on www2.ohchr.org/english/press/hrc/kit/hrcs_fs2.pdf.

490 See section 2.

491 Human rights and access to safe drinking water and sanitation: resolution adopted by the Human Rights Council 2010 4.

492 Donoho. p. 101.

493 Karin Arts, 'Inclusive Sustainable Development: A Human Rights Perspective' (2017) 24 Current Opinion in Environmental Sustainability 58 <http://linkinghub.elsevier.com/retrieve/pii/S1877343517300301>.

494 Sara De Vido, 'The Right to Water as an International Custom : The Implications in Climate Change Adaptation Measures' (2012) 3 Carbon and Climate Law Review 221. p. 225.

3.3.4 THE HUMAN RIGHT TO WATER AS A GENERAL PRINCIPLE OF INTERNATIONAL LAW

In order to establish the human right to water as a general principle of law, proof of its general acceptance and recognition is required.[495] This proof can be based on an analysis of statements as was done in the above-mentioned sections. 'National constitutions can, if read together with soft law such as international declarations, be an expression of a general principle of international law.'[496] It has been demonstrated that consistency is currently lacking regarding the recognition of the human right to water. Only a limited number of states have recognised the right in their constitution or in lower legislation. Therefore, as yet there is not enough evidence for a consistent trend which would justify that the human right to water is regarded as a general principle of international law.[497]

3.4 THE NORMATIVE CONTENT OF THE HUMAN RIGHT TO WATER

3.4.1 INTRODUCTION

As mentioned, General Comment. No. 15 is the most authoritative interpretation of the human right to water. General Comment No. 15 creates a number of consequences even though it is not legally binding.[498] It creates support for the explicit inclusion of the right to water in the ICESCR. General Comment No. 15 is a tool for the CESCR to use when analysing the progress that has been made in realising the enjoyment of the rights in the ICESCR.[499] Elements of the Comment may be included in the dialogue with states. In addition, and most importantly, this Comment helps to define the right to water in relation to its quantity, quality and accessibility. Therefore General Comment No. 15 strongly contributes to the recognition of the right and creates an understanding about the responsibility of governments. It sets standards and guidelines for the parties to the Covenant to follow.[500] The following section discusses the content of the right as explained by General Comment No. 15.

The Comment lays down the human right to water as follows:

> *'the human right to water entitles everyone to sufficient, safe, acceptable, physically accessible and affordable water for personal and domestic uses. An adequate amount of safe water is*

---

495 Simma and Alston. p. 102.

496 Pierre Thielbörger, *The Right(s) to Water The Multi-Level Governance of a Unique Human Right* (Springer Science+ Business Media BV 2014). p. 86.

497 (Winkler, 2012) p. 97.

498 Amy Hardberger, 'Life, Liberty, and the Pursuit of Water : Evaluating Water as a Human Right and the Duties and Obligations It Creates' (2005) 4 Northwestern Journal of International Human Rights 331. p. 349.

499 Tully. p. 43.

500 Cahill. p. 392.

necessary to prevent death from dehydration, to reduce the risk of water-related disease and to provide for consumption, cooking, personal and domestic hygienic requirements.'[501]

3.4.2 PERSONAL AND DOMESTIC USES

The human right to water is not concerned with all water uses.[502] It only covers personal and domestic water uses which ordinarily mean: drinking, personal sanitation, the washing of clothing, food preparation, personal and household hygiene.[503] Any other uses fall outside the scope of this human right.[504] This section will delve into the uses that are covered by the human right to water and will also discuss those which fall outside its scope but do have human rights implications for water allocation and policy. Public policy on water allocation is highly relevant as in the event of water scarcity priorities might have to be set as water is a finite resource. What this could mean is that a state has to decide which uses of water have priority over others, thereby influencing the economy and affecting the environment. What addressing access to water as a human right does is to establish water as an entitlement for individuals when currently 780 million people still do not have access to improved water[505] and the industrial and agricultural sectors continue to use 90%[506] of the world's freshwater resources.

Even though water is essential for life and plays an important part in food security, the right to water does not cover water for agriculture or grazing livestock. Some argue that the scope of the right to water should be expanded so as to also cover access to water for subsistence farming.[507] General Comment No. 15 does not take this approach but does include an advisory provision stating:

'*Attention should be given to ensuring that disadvantaged and marginalized farmers, including women farmers, have equitable access to water and water management systems, including sustainable rain harvesting and irrigation technology. Taking note of the duty in Article 1, paragraph 2, of the Covenant, which provides that a people may not "be deprived of its means of subsistence", States parties should ensure that there is adequate access to water for subsistence farming and for securing the livelihoods of indigenous peoples.*'[508]

---

501 Paragraph 2.

502 Not all domestic uses fall under the protection of the human right to water. Uses such as washing vehicles or filling swimming pools fall outside the sphere of this right. Winkler, *The Human Right to Water Significance, Legal Status and Implications for Water Allocation*.p. 146.

503 (Committee on Economic, Social and Cultural Rights, 2002) par. 12.

504 Misiedjan and Gupta. p. 86 also see Committee on Economic Social and Cultural Rights. par. 6 where priority is given to water for domestic purposes over other water uses.

505 Www.unicef.org/media/files/JMPreport2012.pdf p. 2.

506 Nobonita Chowdhury et al., *The Human Right to Water and the Responsibilities of Businesses: An Analysis of Legal Issues*, University of London 2011 available on: www.ihrb.org/pdf/SOAS-the_Human_Right_to_Water.pdf.

507 See Dinara Ziganshina, 'Rethinking the Concept of the Human Right to Water' (2008) 6 Santa Clara Journal of International Law 113. p. 117.

508 Committee on Economic Social and Cultural Rights. par. 7.

If water for food production would be included under the human right to water it would have far-reaching consequences. Firstly, the quantity of water per person guaranteed by the right would substantially increase. As mentioned, the WHO has established that basic needs can be satisfied with between 20 and 100 l/c/d.[509] However, if one takes into account that it requires 13 litres of water to produce 1 tomato[510], it would be a challenge to decide how much water individuals are entitled to as this will depend on the type of crop which the person is growing and the environmental conditions. For instance, if a person is producing water-intensive crops in an arid climate, they will need high quantities of water.[511] It is then the question of whether or not it is responsible to provide access to water under the human right to water for this type of subsistence farming.

Secondly, the quality standards for water for domestic purposes and those for household purposes differ. And these water uses also depend on different infrastructures.[512] Thirdly, another challenge would be to distinguish between subsistence farming and agriculture in general. Agriculture not only covers subsistence farming but also small-scale agriculture and the raising of livestock.[513]

Lastly, one of the major contributions that the human right to water offers is that it aims to influence water management by giving priority in water allocation to personal and domestic uses for all.[514] According to the UNDP personal and domestic uses only make up a tiny fraction of the total water use, between 5 and 10 percent. Agriculture, on the other hand, accounts for more than 80 percent in developing countries. Including access to water for agriculture within the human right to water would increase competition between the two uses and eliminate the counterbalance of prioritising water for personal and domestic uses which is the objective of the human right to water in its current form.

These considerations illustrate why water for food production should not be covered by the human right to water as articulated under the ICESCR and why separating access to water for food cultivation and access to water for domestic needs is the most appropriate approach. In addition, access to water for agriculture is already provided under the right to adequate food.

Another use which falls outside of the scope of the human right to water is access to water for environmental protection and to sustain ecosystems.[515] When mentioning different water uses, water for ecosystems is often completely ignored, possibly because it is only indirectly related to human needs. Using water for domestic and personal use, agriculture and industry are mostly mentioned. Yet, water for ecosystems is of significant importance. A way in which to secure the amount of water needed to sustain

509 G Howard and J Bartram, 'Domestic Water Quantity, Service Level and Health' (2002). p. 22.

510 Marta Antonelli and Francesca Greco, *The Water We Eat* (Marta Antonelli and Francesca Greco eds, Springer 2015). p. 6; concerning the concept of virtual water see J.A. Allan, *'Virtual Water': A Long Term Solution for Water Short Middle Eastern Economies?*

511 (Antonelli & Greco, 2015) p. 8.

512 Winkler, *The Human Right to Water Significance, Legal Status and Implications for Water Allocation*. p. 129.

513 (Winkler, 2012) p. 29.

514 See (Winkler, 2012).

515 OHCHR, *Fact Sheet No. 35, The Right to Water* (2010) <www.ohchr.org/Documents/Publications/FactSheet35en.pdf>. p. 12.

ecosystems could be by determining the minimum environmental flow. Chapter 4 will explore this topic further.

3.4.3 AVAILABILITY, QUALITY AND ACCESSIBILITY

Paragraph 12 of General Comment No. 15 contains the substantive standards of the human right to water. These include the three principles of availability, quality and accessibility which the Committee often refers to.[516] According to General Comment No. 15 each person should have a sufficient amount of water available to them.[517] They should be able to use this water to prevent death from dehydration and disease. They should also be able to prepare food and use water for hygienic purposes. The UN has not determined how much water is a sufficient amount for domestic and personal use. Nevertheless, the CESCR follows the WHO guidelines which state that 20 litres per day could be the minimum not excluding that some might need additional water due to health, climate or working conditions.[518] Therefore local conditions can be an argument for adjusting the amount which is allowed by the human rights framework.[519] 'Both the law and national courts and science increasingly agree that a person requires 50 lcd (litres per capita per day) to meet their basic human needs including drinking, washing, food preparation and personal hygiene. This means that any use of water above that level is beyond the basic human need and hence not protected by human rights law.'[520]

Apart from having sufficient water available, which is an issue of quantity, the Committee has also identified certain standards for the quality of water.[521] To meet the quality standards identified by the Committee water should be clean, free from harmful substances like micro-organisms and chemicals and suitable for domestic and personal use.[522] This requirement is applicable to all sources of water provision. Persons may expect this quality when retrieving water through their pipes, when retrieving it from water tankers, vendors and (protected) wells.[523]

Thirdly, water and water facilities must be accessible to everyone, without discrimination, within the jurisdiction of the state. Accessibility refers to the physical and affordable accessibility of safe water on a non-discriminatory basis and access to information related thereto.[524] In all spheres of life, water should be accessible to

516 See for instance General Comment No. 14 The right to the highest attainable standard of health.
517 Paragraph 12.
518 Report of the Independent Expert on the issue of human rights obligations related to access to safe drinking water and sanitation, U.N. Doc.A/HRC/12/24 (2009), paragraph 12.
519 General Comment No. 15, note 33.
520 Melvin Woodhouse and Malcolm Langford, 'There Is No Human Right to Water for Livelihoods' (2009) 28 Waterlines 5 <http://developmentbookshelf.com/openurl.asp?genre=article&id=doi:10.3362/1756-3488.2009.002> accessed 6 March 2013. p. 5 – 12.
521 Committee on Economic Social and Cultural Rights. par.12.
522 Committee on Economic Social and Cultural Rights, 2003) par. 72–74.
523 (OHCHR, 2010) p. 9.
524 Paragraph 12(c).

all. This includes groups which requires special attention like children, the elderly, the disabled but also the homeless and those living in institutions such as prisons or psychiatric hospitals. Safe water should be available during every aspect of person's life: the home, workplace, school, and other public institutions.[525] And collecting water without an unreasonably long wait for the water is also a requirement.

Water should be affordable to everyone and it should not interfere with the realisation of other human rights. For instance, if a family spend most of their income on water and are not able to afford education for the children in the family, there is a conflict in realising two different rights. Citizens should not have to choose between realising their right to appropriate housing or their right to water and they should be able to acquire basic goods and education as well. This does not mean, however, that states should make water universally available and free of charge.[526] The Committee has chosen a different perspective from, for instance, that of Agenda 21, where it was stated that water for basic needs should be free of charge.[527] A reason for this is that the Committee has chosen for a comprehensive approach by taking the whole of the Covenant into account. The Covenant already finds that individuals have the right to a sufficient income in order to meet their basic needs and realise their human rights.[528]

Water services should be affordable irrespective of who supplies the services. States are obliged to adopt regulatory frameworks that make water progressively available, accessible and affordable to all.[529] Private parties and/or the public sector can provide these services. Implementing an effective tariff structure which allows low income households to pay a lower rate for basic amounts of water would be a possible solution to address the affordability of water for these households.[530] The criterion of affordability is broad, flexible and adaptable, which gives states the opportunity to apply situation-specific solutions. In human rights different approaches can be implemented that address the issue looking at a particular situation from the individual's perspective.

3.5 STATE OBLIGATIONS REGARDING THE HUMAN RIGHT TO WATER

From the previous sections it can be concluded that there is an independent self-standing human right to water. The following section gives an overview of the state

525 (Committee on Economic, Social and Cultural Rights, 2003) p. 6.
526 (Committee on Economic, Social and Cultural Rights, 2002) p. 6.
527 Ibid., paragraph 27.
528 Malcolm; Langford, 'Ambition That Overleaps Itself? A Response to Stephen Tully's Critique of the General Comment on the Right to Water' (2006) 24 Netherlands Quarterly of Human Rights 433. p. 277.
529 Paragraph 12(c)2.
530 De Albuquerque and Roaf, *On The Right Track. Good Practices in Realising the Rights to Water and Sanitation*. p. 35.
Those who cannot depend on the public infrastructure and as a result buy water from, for instance, water trucks pay disproportionately high tariffs.

obligations which arise from the human right to water. Firstly, the core content of the right will be discussed. This will be followed by discussing the obligation to realise the right to water on a progressive basis. Then this section will discuss the three obligations to respect, protect and fulfil the right and this will be followed and concluded by the obligation of non-discrimination.

3.5.1 CORE CONTENT

According to General Comment No. 3 the rights in the ICESCR have minimum core obligations which ensure that the minimum essential level of each of the rights are satisfied.[531] This minimum core of rights related to the essential prerequisites for leading a life with dignity. States should guarantee the minimum core unless this is impossible due to circumstances which lie outside of the control of the state such as armed conflict and constraints in resources.[532] The human right to water also has such a core of rights. General Comment No. 15 formulates a number of aspects which the Committee considers as core obligations which are of immediate effect. The list is non-exhaustive, therefore more elements can be included. Firstly, states are obliged '[t]o ensure access to the minimum essential amount of water, that is sufficient and safe for personal and domestic uses to prevent disease.'[533] Secondly, the principle of non-discrimination is seen as part of the core. Water facilities and services need to be provided on this basis, especially when supplying disadvantaged and marginalised groups. The remainder of the obligations concern the monitoring of the realisation of the right, guaranteeing personal security and engaging in disease control. With these core obligations the Committee also emphasises that both states and other relevant parties should cooperate in order to realise the right. They should also be willing to support developing states to fulfil the right for those within the territory of these developing states, by for instance technical and economic assistance.

3.5.2 PROGRESSIVE REALISATION

States should be 'achieving progressively the full realisation of the rights recognised'[534] in the Covenant. Article 2(1) of the Covenant explains this by stating: 'Each State Party to the present Covenant undertakes to take steps, individually and through international assistance and co-operation, especially economic and technical, to the maximum of its available resources, with a view to achieving progressively the full realization of the rights recognized in the present Covenant by all appropriate means, including particularly the adoption of legislative measures'.

531 (Committee on Economic, Social and Cultural Rights, 1990) par. 37.
532 Riedel, Giacca and Golay, 'The Development of Economic, Social, and Cultural Rights in International Law'. p. 14.
533 (Committee on Economic, Social and Cultural Rights, 2002) p.37.
534 Article 2(1) ICESCR.

This obligation for a state to make use of its maximum available resources with the goal of progressively realising rights is particularly relevant in the case of the human right to water for several reasons.[535] Firstly, this obligation is mostly interpreted from an economic perspective where states are required to use their financial resources to meet this obligation.[536] However, especially in relation to the human right to water not only the use of financial resources is relevant but also the natural resource water is relevant. The obligation to use maximum resources could be interpreted to mean that states are required to make use of the maximum amount of available water resources. However, because it concerns a natural resource states should be required to make sustainable use of their water resources. In order to use water resources sustainably, states should take into account the environmental flow of rivers and the recharging capacity of groundwater resources.[537] Certain water resources can be classified as non-renewable as they can be depleted within a lifetime.[538] In this light, states should be required to take environmental aspects into account when realising the human right to water.[539]

This obligation also relates to the principle of non-retrogression which is connected to the obligation of progressive realisation. This principle entails that states are prohibited from taking deliberate retrograde steps which cannot be justified by for instance severe economic hardship or force majeure.[540] The obligation of progressive realisation and the principle of non-retrogression should allow analysing particular national contexts. Thereby leaving room for the different capacities of states.[541]

With regard to water this could mean that retrograde steps by states can be justified in worse-case scenarios. Such situations can for example occur when they are due to climate change or drought resulting in a situation where water availability has decreased. The question that emerges under these circumstances is whether the state has played a part in the water scarcity due to certain decisions it has made in its water allocation policy. The obligation of progressive realisation could for this reason also have as a consequence that states are obliged to justify any decisions they have made in relation to water allocation. If, for instance, a state has decided to allocate more water resources to industry while overlooking the need for the provision of drinking water for its citizens, the Committee could be of the opinion that the state has failed to meet its obligations.[542] Using sustainability as an instrument could provide a solution.

---

535 (Committee on Economic, Social and Cultural Rights, 1990) par. 2.

536 Robert E Robertson, 'Measuring State Compliance with the Obligation to Devote the " Maximum Available Resources " to Realizing Economic, Social, and Cultural Rights' (1994) 16 Human Rights Quarterly 693. p. 15.

537 Stephen C Mccaffrey, 'Keynote: Sustainability and Sovereignty in the 21th Century' (2013) 41 Denver Journal of International Law and Policy 507. p. 512.

538 Antonelli and Greco. p. 7.

539 Christopher Jefferords and Farhed Shah, 'On the Natural and Economic Difficulties to Fulfilling the Human Right to Water Within a Neoclassical Economics Framework' (2012). p. 4.

540 Riedel, Giacca and Golay, 'The Development of Economic, Social, and Cultural Rights in International Law'. p. 15. Chapter 5 elaborates more on this topic.

541 Ibid.

542 Riedel, Giacca and Golay, *Economic, Social, and Cultural Rights in International Law.*

Interestingly, the Committee on Economic, Social and Cultural Rights has affirmed sustainability is an requirement for the realisation of the human right to water.[543]

Secondly, Article 2(1) explains that in order to fulfil the rights of the Covenant states should request international cooperation and assistance if this is necessary. Consequently, those states which are unable to fulfil their obligations due to resource restraints are still required under this article to seek assistance from other states. Mostly, this refers to the supportive capacity of developing states. Nevertheless, in the case of water scarcity also lesser developed states possessing adequate water resources could provide this supportive capacity. States which have supportive capacity have two obligations in this sense. They are required to progressively realise the human right to water for their citizens but also to render assistance to other states which are unable to do so.[544]

3.5.3 OBLIGATION TO RESPECT, PROTECT AND FULFIL

In addition to the obligation to progressively realise the human right to water, states also have the obligation to (a) respect, (b) protect and (c) fulfil with regard to the fulfilment of the human right to water.[545]

a) States are prohibited from interfering directly or indirectly with the enjoyment of this right. They may not take any measures that violate individuals' right to water within their jurisdiction. An example of a violation is limiting local communities' access to traditional water sources. Such acts can be identified as acts of commission.[546]
b) States should protect their citizens from infringements by third parties. This obligation, unlike the obligation to fulfil and respect, is indirect. Here the state needs to ensure that others do not violate rights to which it is committed.[547] This can be done by implementing prohibitive legislation for human rights abuses by third parties. The human right to water is only applicable to the relationship between the state and the citizen. Violations of the right can only be attributed to states even if they concern third parties.[548] If a state has negated to adopt these measures this constitutes an act of omission.[549]
c) States are required to facilitate, promote and provide for the rights of individuals and other positive obligations. The state needs to actively pursue the realisation of the right to water for individuals. The realisation can be done through legislation,

543 (Committee on Economic, Social and Cultural Rights, 2002) para. 11.
544 Riedel, Giacca and Golay, 'The Development of Economic, Social, and Cultural Rights in International Law'. p. 15.
545 General Comment No. 15, paragraph 20.
546 Http://untreaty.un.org/ilc/texts/instruments/english/draft%20articles/9_6_2001.pdf.
547 Ibid. p. 19.
548 Even though private parties are required to live up to human rights standards.
549 Chowdhury and others. p. 8.

the judiciary, the administration and other measures.[550] It also means that the state is obliged to provide what is necessary for the enjoyment of this right if groups or individuals are unable to provide for it themselves. In the case of the human right to water states are obliged to make water accessible at a reasonable price for those who cannot afford it or even free of charge if necessary.

The conduct of corporations may come into conflict with the requirements under the human right to water in several situations. The most common situations are in the context of private sector participation in water services and in the event of water pollution or depletion.[551] The "Protect, Respect, Remedy" Framework for Business and Human Rights (the Ruggie Framework)[552] and its UN Guiding Principles[553] set authoritative standards for businesses to address and prevent negative business-related human rights impacts. The three pillars of this framework are:

1. the state's duty to protect against human rights abuses by third parties, including business;
2. the corporate responsibility to respect human rights; and
3. greater access by victims to an effective remedy, both judicial and non-judicial.

Because of the increasing water crisis, it is essential for companies to align their practices with the human right to water and sanitation. And states are obliged to oversee the actions of corporations.[554]

Besides obligations which apply on the state's territory, states also have the obligation to respect the enjoyment of this right in all other states.[555] This means that the state and third parties within their jurisdictional control are prohibited from interfering with the enjoyment of the right. States can realise this by concluding agreements on water issues for development and encouraging companies to implement due diligence measures.[556] In this respect states may for example require companies to guarantee that their production does not make use of scare water resources in foreign states.[557]

---

550 (Committee on Economic, Social and Cultural Rights, 1990) par. 5.

551 (Winkler, 2012) p. 277.

552 Https://business-humanrights.org/sites/default/files/reports-and-materials/Ruggie-report-7-Apr-2008.pdf.

553 United Nations, 'Guiding Principles on Business and Human Rights Guiding Principles on Business and Human Rights' 35.

554 Thielbörger, *The Right(s) to Water The Multi-Level Governance of a Unique Human Right.*

555 (Committee on Economic, Social and Cultural Rights, 1990) par. 18; See Takele Soboka Bulto, *The Extraterritorial Application of the Human Right to Water in Africa* (Cambridge University Press 2014).

556 Ibid.

557 Pierre Thielbörger, *The Current Legal Status of the Right to Water* (Springer Berlin Heidelberg 2014) <http://link.springer.com/10.1007/978-3-642-33908-0> accessed 12 March 2014. p. 74 & see MA Robesin, 'Katoen En Waterschaarste', *Duurzame handel in juridisch perspectief* (2014).

3.5.4 NON-DISCRIMINATION[558]

The principle of non-discrimination, laid down in Article 2(2), is of great importance for the implementation of the Covenant and the human right to water. It states that: '[t]he States Parties to the present Covenant undertake to guarantee that the rights enunciated in the present Covenant will be exercised without discrimination of any kind as to race, colour, sex, language, religion, political or other opinion, national or social origin, property, birth or other status.'

This principle is part of the core elements of the Covenant and requires immediate application and is binding for all levels of government and for all state entities.[559] Rights contained in the Covenant are often resource-dependent. This principle can however be immediately implemented as it usually only requires governmental commitment.[560] 'Even the least developed countries do not need to discriminate against the most marginalized and disadvantaged individuals and groups of society when legislating new norms, or applying and interpreting existing legislation.'[561]

General Comment No. 20 explains that the principle deals with both formal and substantive discrimination. As a consequence, states should not only remove discriminatory provisions on prohibited grounds from their constitution, laws and policies[562] but should also prevent discrimination in practice.[563] Next to this, states should be aware of direct discrimination where an individual is treated unequally when compared to another person and this is merely based on a reason related to a prohibited ground and they should also be aware of indirect discrimination where individuals, for reasons related to a prohibited ground, are disproportionality negatively affected by a certain situation.[564]

With regard to the human right to water, the principle of non-discrimination can be an effective vehicle for tackling the lack of access to water for vulnerable groups. Often the justiciability of this principle is found challenging. Nevertheless, in several national cases concerning the human right to water the principle of non-discrimination[565] has been put on the forefront. For instance, in *Kennedy v City of Zanesville*[566], the only predominately black neighbourhood in the area[567] was denied

558 (COMMITTEE ON ECONOMIC, SOCIAL AND CULTURAL RIGHTS, 2002) PAR. 12.

559 Catarina De Albuquerque and Virginia Roaf, 'Principles', *Realising the human rights to water and sanitation: A handbook by the UN Special Rapporteur Catarina De Albuquerque* (2014). p. 10.

560 Riedel, Giacca and Golay, 'The Development of Economic, Social, and Cultural Rights in International Law'. p. 16.

561 Ibid. p. 16.

562 International Covenant on Economic, Social, and Cultural Rights 2009 1. par. 8.

563 (CESCR, 2009) par. 8.

564 (CESCR, 2009) par. 10.

565 See for instance: Lindiwe Mazibuko & ors v. the City of Johannesburg & ors, Case no. 06/13885, High Court of South Africa (Witwatersrand Local Division), Judgment of 30 April 2008; Abadallah Abu Massad and others v Water Commissioner and Israel Lands Administration, Supreme Court Civil Appeal 9535/06, Judgement of 5 June 2011.

566 Kennedy v City of Zanesville Southern District Court of Ohio, Eastern Division 10 July 2008.

567 City of Zanesville, Ohio, United Stated of America.

connection to the water supply system for over 50 years which infringed on their human right to water. This neighbourhood also endured an additional challenge as they could not depend on groundwater as it was contaminated due to mining activities and therefore unacceptable for domestic purposes due to the mining activities. The element of discrimination was proven by comparing the access to water of this lone black neighbourhood to the surrounding white neighbourhoods. The District Court concluded that 'the discriminatory, regular, and continuous decisions to pass over Coal Run and bring water to predominantly white areas despite being on notice of the Coal Run neighbourhoods' need for water'[568] constituted a violation of the fair housing laws.

Chapter 2 explained that vulnerability in relation to the human right to water should not only be based on the grounds of discrimination and other social constructed circumstances. perspective of environmental vulnerability should be included. With regard to realising the human right to water, equitable water management is required as part of the principle of non-discrimination. This will assist in bringing about accessibility and affordability.[569]

3.6 JUSTICIABILITY OF THE HUMAN RIGHT TO WATER

The justiciability of the human right to water[570] similar to other economic, social and cultural rights is often questioned.[571] Literature mentioned several reasons for the human right to water not being susceptible to judicial determination. Firstly, the content of socio-economic rights like the human right to water is often seen as unclear. This lack of clarity is however not a characteristic of socio-economic rights but caused by its exclusion from adjudication.[572] The abovementioned sections demonstrate the right has been invoked in national courts and interpreted by judges. These examples demonstrate that the judiciary assists in specifying the content of the right.[573] This contributes to tackling national and local challenges in realising the human right to water for rights-holders and adds to the justiciability of the right.

568 Kennedy v City of Zanesville Southern District Court of Ohio, Eastern Division 10 July 2008 p. 70.

569 Cahill. p. 404.

570 Selection: Anonymous.; Langford, 'Ambition That Overleaps Itself? A Response to Stephen Tully's Critique of the General Comment on the Right to Water'; Ramin Pejan, 'The Right to Water: The Road to Justiciability' (2004) 36 The George Washington International Law Review 1181; Michael J Dennis and David P Stewart, 'Justiciability of Economic, Social, and Cultural Rights : Should There Be an International Complaints Mechanism to Adjudicate the Rights to Food, Water, Housing, and Health ?' (2012) 98 The American Journal of International Law 462; Winkler, 'Judicial Enforcement of the Human Right to Water – Case Law from South Africa, Argentina and India'.

571 Fons Coomans, *Justiciability of Economic and Social Rights: Experiences from Domestic Systems* (Intersentia 2006).

572 See Marius Pieterse, 'Coming to Terms with Judicial Enforcement of Socio-Economic Rights' (2004) 20 South African Journal on Human Rights 383.

573 Winkler, *The Human Right to Water*.p. 237.

Secondly, justiciability of the human right to water is questioned in relation to the separation of powers.[574] With the enforcement of socio-economic rights, courts are often confronted with resource questions.[575] It is not within the power of the judiciary to decide how governments should allocate their resources not is the court allowed to rearrange budgets and adapt policies. The role of the court, in examining resources questions around the enforcement of socio-economic rights, is limited to setting boundaries for the benefit of meeting human rights obligations.[576] The court does not go beyond its tasks. Thirdly, the capacity of judges to appropriately examine is used as a reason for the non-justiciability of the human right to water. Judges are often asked to deliberate on issues they do not have expertise on. This reasoning would therefore possibly influence the extent to which and how the court will examine the case and not whether they have jurisdiction.[577] From all this it can be concluded that the human right to water is not characterised by non-justiciability and that judicial bodies contribute to articulating its content.

3.7 CONCLUSION

Table 3.2 Overview of the water uses covered by the human right to water in relation to its legal foundation

Legal foundation	Drinking	Hygiene	subsistence farming	Other uses Specify
Right to life (Article 6 ICCPR)	x			
Right to be free from cruel, inhumane, degrading treatment (Article 7 ICCPR)	x	x		
Right to food (Article 11 ICESCR)	x		x	
Right to housing (Article 11 ICESCR)	x	x		
Right to health (Article 12 ICESCR)	x	x		
Right to sanitation (Articles 11 & 12 ICESCR)		x		
Right to a healthy environment (Article 12 ICESCR)	x	x	x	x
Right to partake in culture life (Article 15 ICESCR)				x
Right to development	x	x	x	x

The human right to water is not included in the catalogue of human rights. This chapter has therefore examined the legal foundations. However, it firstly discussed how the

574 Carlos Bernal, 'The Right to Water: Constitutional Perspectives from the Global South', *International Environmental Law and the Global South* (2015) p. 280.

575 Lucy A Williams, 'Resource Questions in Social and Economic Rights Enforcement: A Preliminary View', *Social and Economic Rights in Theory and Practice: A Critical Inquiries* (2015).

576 Pejan p.1183.

577 Winkler, *The Human Right to Water*.p. 232.

human right to water has emerged as a human right. It was established that the right was first formulated in soft environmental law instruments and therefore did not create entitlements for individuals. The right was also firmly placed in the overall setting of water management. In 2002, the Committee on Economic, Social and Cultural Rights took the initiative to interpret the right as a self-standing human right. This legal recognition of the right was followed by political recognition by the UNGA and UNHRC.

Also, the number of states which have included a human right to water in legislation such as constitutions or lower level laws has increased over the years. In addition, several constitutional courts have used their national constitutions to read the human right to water into it. Resulting in the human right to water becoming a part of a growing state practice. 'What is understood to be a human right to water is increasingly recognised by nations and is resulting in a consistent practical picture of what the right really is in the eyes of the law.'[578] This development could lead to the recognition of the human right to water as customary law.

Even though the right is legally and politically recognised there are unresolved questions concerning its legal foundations. In order to include access to water as a human right, this must be implicitly deduced from treaty provisions or be recognised as part of customary law or general principles of law. This chapter has analysed these sources of law and has concluded that there are several provisions within international human rights law from which a human right to water can be derived as the figure above shows.

Water is fundamental to all human rights as by guaranteeing access to water the fulfilment of other human rights will also be supported. Water is essential to realising other fundamental human rights and lies at the heart of the right to life which is protected under Article 6 ICCPR.[579] But also Article 7 ICCPR supports access to water at the survival level, particularly when individuals are dependent on states such as detainees. The benefits of recognising the human right to water under the ICCPR include the wide support of the Covenant and far-reaching state obligations. However, the right has limited coverage as it is only concerned with protecting individuals from death by dehydration.

Provisions under the ICESCR provide less narrow interpretations of the human right to water. Most authoritatively, the right has been derived from Article 11 on an adequate standard of living and is connected to the right to health and adequate housing by the CESCR in General Comment No. 15. This Comment also assists in clarifying the content of the right. It states that 'everyone has the right to sufficient, safe and acceptable, physically accessible and affordable water for domestic and personal uses.' The right obligates states to respect, protect and fulfil the right. It also requires states to realise its core content under all circumstances and to strive for progressive realisation. From the analysis in this chapter it can be concluded that the human right to water is a self-standing independent right and its content is still under development as it is not explicitly mentioned in universal human rights treaties.

---

578 Woodhouse and Langford. p. 5 – 12.

579 John Scanlon, Angela Cassar and Noemi Nemes, *Water as a Human Right?* (IUCN 2004). p. 18.

CHAPTER 4
SUSTAINABILITY IN THE CONTEXT OF WATER

'Fresh water is a finite and vulnerable resource, essential to sustain life, development and the environment.'[580]

4.1 INTRODUCTION

The previous chapter laid out the legal foundations and content of the human right to water. Its realisation, as previously explained, depends on the availability and allocation of water resources. Ensuring vulnerable groups' sustainable access to water for domestic purposes has several aspects in relation to water resources management which should be considered. It is often argued that providing access to those who currently do not have such access to water will increase environmental degradation and the depletion of water resources.[581] However, as mentioned, on a global basis household consumption accounts for less than 10 percent of the overall human water use. Because of this, household consumption does not pose the biggest threat to water resources.[582] Also, there is sufficient water to meet the basic needs of all. Even in countries where water is limited, there is still enough to meet basic needs.[583] The issue of who obtains water and how much is not exclusively an issue of the availability of water resources. As the Human Development Report explains: 'the scarcity at the heart of the global water crisis is rooted in power, poverty and inequality, not in physical availability'.[584] Ensuring

---

580 *Principle 1 of the Dublin Statement on Water and Sustainable Development.*

581 See for instance Trigueros. where it is argued that: To answer the questions posed in the introduction: the considerably weakened state of international environmental law, following the shift in focus from environmental protection to a right to development, coupled with the potential realization of a human right to water could very well lead to rapid depletion of the planet's freshwater resources. (p. 625)

582 Bourquain. p. 3.

583 Winkler, *The Human Right to Water Significance, Legal Status and Implications for Water Allocation.* p. 7; (Khalfan et al., 2008) p. 87.
According to Ashfaq Khalfan and others, *Manual on the Right to Water and Sanitation* (2008).: In only in a few regions of a few countries would there be insufficient water resources for essential domestic purposes such as drinking, washing (clothing and personal hygiene), cooking and sanitation – if such domestic uses were prioritised over other uses.

584 United Nations Development Programme. p.2.

access to water for vulnerable groups is therefore also a question of prioritisation in allocation.

This does not mean, however, that the quality and quantity of freshwater resources should not be monitored. Additionally, the state of freshwater resources and the ecosystems which purify and replenish them are an essential element in realising the human right to water. Overuse of water and the contamination of water resources are among the challenges which cause water scarcity. This can result in people losing access to traditional water sources, simple water filtering services becoming insufficient and the depletion of water sources.[585] All these issues make the availability of sufficient and safe water problematic. In other words, securing the right to water for those who are currently unserved and future generations demands sustainable water resources management and the prioritisation of allocating water for domestic purposes.[586]

This perspective is also supported by General Comment No. 15 which mentions: 'the manner of the realisation of the right to water must also be sustainable, ensuring that the right can be realised for present and future generations.'[587] Even though the Committee mentions sustainable access on several occasions in this Comment[588], it does not elaborate on what sustainable access entails, but it does refer to definitions of sustainability and sustainable development as discussed in the Rio Declaration.[589] In 2013, the Special Rapporteur on the human right to safe drinking water and sanitation offered more insight into the relevance of sustainability for the human right to water. In this report, she argues for sustainability as a principle of human rights.[590] Her report primarily deals with economic considerations such as budgeting and economic growth and the connection between sustainability and core concepts of human rights.[591] It nevertheless still leaves a need for deliberation on environmental sustainability and sustainable development in relation to the human right to water.

These two documents, General Comment No. 15 and the report of the special rapporteur, were not adopted by states. Therefore it is difficult to argue that the inclusion of sustainability as part of the human right to water is supported by states and

585 Khalfan and others. p. 86.

586 G Björklund and J Sjödin, *The Human Right to Water and Sanitation Securing Access to Water for Basic Needs* (2010). p. 4; Khalfan and others. p. 87.

587 Committee on Economic Social and Cultural Rights. par. 11.

588 See ibid. paragraphs 26, 28,

589 See footnote 5. The Committee refers to the Rio Declaration and Agenda 21 for a definition of sustainability. Footnote 11: 'For a definition of sustainability, see the Report of the United Nations Conference on Environment and Development, Rio de Janeiro, 3–14 1992, Declaration on Environment and Development, principles 1, 8, 9, 10, 12 and 15; and Agenda 21, in particular principles 5.3, 7.27, 7.28, 7.35, 7.39, 7.41, 18.3, 18.8, 18.35, 18.40, 18.48, 18.50, 18.59 and 18.68.'
Footnote 22: Agenda 21, chaps. 5,7 and 18; and the World Summit on Sustainable Development, Plan of Implementation (2002), paras. 6 (a), (l) and (m), 7, 36 and 38.
The Committee mentions sustainable access to water in paragraphs: 11; 26; 28;

590 De Albuquerque, 'Report of the Special Rapporteur on the Human Right to Safe Drinking Water and Sanitation'. paragraph 9.

591 Ibid.
The considerations of the Special Rapporteur will be discussed in the following chapter.

will become part of their implementation efforts. However, support is increasing. With the development of the Sustainable Development Goals in 2015, states did adopt Goal no. 6 on ensuring access to water and sanitation for all with one of the targets ensuring sustainable withdrawals and the supply of freshwater.[592]

As mentioned, the realisation of the human right to water for vulnerable groups in relation to the management of water resources demands a reallocation of water for domestic purposes[593] and the protection and sustainable use of water resources. This chapter looks for support in soft and hard law instruments of international environmental law and international water law which assist states in shaping efforts aimed at the sustainable management of water resources which will facilitate the realisation of vulnerable groups' sustainable access to water for domestic purposes. It identifies the most relevant aspects of these bodies of law and analyses them in relation to water resources.

This study uses the references to sustainability and sustainable development in General Comment No. 15 as a point of departure. General Comment No. 15 makes a strong reference to instruments such as: Agenda 21, the Dublin Statement on Water and Sustainable Development, the Plan of Implementation of the World Summit on Sustainable Development and the Rio Declaration on Environment and Development. The Comment also refers to instruments which directly relate to the allocation of water resources, such as the UN Convention on the Law of Non-Navigational Uses of Watercourses. This study analyses these instruments and where it can assist in explaining certain principles, for instance, it makes use of additional provisions from other instruments and relevant jurisprudence.

The analysis of sustainability in the context of water lies behind the genesis and early articulation of sustainability and sustainable development within international law and policy. I will discuss the timeline of sustainable development and the changes it has gone through. In section 4.3 the legal status of the concept of sustainable development will be discussed. This is followed by a discussion of international norms which relate to freshwater resources. The principles which operationalise sustainable development are laid out in section 4.5, while section 4.6 contains a conclusion.

4.2 GENESIS AND DEVELOPMENT OF SUSTAINABILITY AND SUSTAINABLE DEVELOPMENT IN INTERNATIONAL LAW AND POLICY

Sustainability and sustainable development are both elusive terms which are widely uses across themes and disciplines. This section uses sources of international law[594]

592 Sustainable Development Goals, goal no. 6.

593 The prioritization of water for domestic purposes will be discussed in chapter 5.

594 According to Article 38(1) of the Statute of the International Court of Justice there are four sources of international law, namely: (a) international conventions, (b) international custom, (c) general principles and (d) jurisprudence.

and international environmental policy to determine how these terms have emerged within the frameworks of international environmental law and policy. What can be observed is that during their development these terms have been understood to include different interests, aspects and values which have occupied central stage depending on the political and environmental climate at the time the specific instrument was formulated.[595] The standards set in different environmental law instruments reflect the environmental and developmental concerns of the times in which they were developed as well as the standards of the context to which they apply.[596] What can also be observed is the increasing influence of science.[597] Developments in environmental science often coincided with or were followed by environmental law.[598]

A large part of the development of sustainability and sustainable development in international law has taken place within international environmental law. This branch of public international law deals with a large variety of subjects such as: environmental problems in relation to economic development, the relationship between developed and developing states and the relationship between science, ecology and policy.[599] It aims to address specific global environmental problems including the loss of biodiversity, atmospheric pollution and the management of freshwater resources. The concepts of sustainability and sustainable development have emerged as central concepts in debates on tackling these problems. The following sections show that sustainability is an early conceptualisation of the idea of managing natural resources for the benefit of current and future generations. However, the idea later evolved into sustainable development which became the dominant term in international environmental law and policy.

The notion of sustainability has a long tradition stretching back to the Age of Enlightenment in Western society and pre-colonial times in the Global South.[600] For instance, the principle of sustainability was introduced through laws in Europe, specifically applying to forests. Article 2 of the Bavarian Forest Act of 28 March 1852 read: 'The management of state-owned forests has to follow sustainability as its highest principle'. Another example of sustainability as state practice can be found in the1893 *Pacific Fur Seals Arbitration* case[601] where the United States advocated the need to prevent unsustainable exploitation in relation to fur seals in the Pacific Ocean.[602] From

595 Klaus Bosselmann, *The Principles of Sustainability* (Ashgate Publishing Limited 2008). p. 25.

596 Winston Anderson, *Principles of Caribbean Environmental Law* (Environmental Law Institute 2012). p. 46.

597 Refer to building block: water systems knowledge and technology.

598 See for instance Aðalheiður Jóhannsdóttir, Ian Cresswell and Peter Bridgewater, 'The Current Framework for International Governance of Biodiversity : Is It Doing More Harm Than Good ?' (2010) 19 Review of European Community and International Environmental Law 139. on the science and policy interface relating to the Rio conventions.

599 Anderson. p. 19.

600 See for instance the Seventh Generation philosophy of the Iroquois Confederations mentioned in paragraph ... on intergenerational equity. And the Maori of New Zealand www.firstfound.org/david%20miller.htm.

601 John Basset Moore, I International arbitrations p. 763–64 (1898). Bering Sea Arbitration of 1893.

602 Alhaji BM Marong, 'From Rio to Johannesburg: Reflections on the Role of International Legal Norms in Sustainable Development' (2003) 16 Georgetown International Environmental Law Review 21. p. 3.

these examples and others, it can be concluded that sustainability was first applied as ecological sustainability.

One of the primary notions within international law is the doctrine of the sovereignty and equality of states.[603] In principle all states enjoy an exclusive right over natural resources within their territorial limits.[604] From this it can be assumed that states are allowed to strive for economic development at the expense of the environment. For a long time during industrialisation and during colonialism this had been the direction followed by national policies. However, with environmental problems occurring questions arose on how to deal with these challenges especially focusing on the environment in relation to economic development.[605] As these problems were complex and global in nature, it was deemed necessary to analyse and deal with them in a global context and international law was seen as potentially being able to provide such a platform.[606]

The idea that the environment determines the confines for development was first presented and summarised in the widely acknowledged report '*The Limits of Growth*' published by the Club of Rome in 1972. Here it was explained that, amongst other things, population, food production, the consumption of natural resources and pollution were growing exponentially.[607] The Club of Rome considered economic growth to be heading for a collision with ecological sustainability and argued for altering growth trends on the basis of economic and ecological stability.[608] At the same time the UN established its United Nations Environment Programme (UNEP) in 1972 with the view that economic growth and ecological sustainability could be reconciled. These two perspectives became dominant within the international discourse on sustainability. On the one hand, there was the perspective that ecological sustainability needed to set the boundaries for economic development and, on the other, the perspective that a balance should be sought between the environment and social and economic goals.[609] This competition between the two perspectives was won by the latter and international environmental law quickly followed this chain of thought.

International environmental law, like any other branch of public international law, holds on to the well-established structures within and linkages between states, economic growth and international law.[610] Whether growth is criticised and ecological sustainability is preferred, or growth is validated and environmental sustainability is seen as being equal to economic growth and social justice, is important for how

603 Anderson. p. 20; Marc Bungenberg and Stephan Hobe, 'Permanent Sovereignty over Natural Resources' [2015] Permanent Sovereignty Over Natural Resources 1. p. 10.

604 See Mccaffrey, 'Keynote: Sustainability and Sovereignty in the 21th Century'.; Onzivu. p. 877.

605 Bosselmann. p. 153 Kiefer and Brölmann. p. 183.

606 Ibid. p. 183.

607 Kerri Woods, *Human Rights and Environmental Sustainability* (Edward Elgar Publishing 2010). p. 121.

608 Donella H. Meadows et al., The Limits to Growth available on www.donellameadows.org/wp-content/userfiles/Limits-to-Growth-digital-scan-version.pdf last accessed October, 2016.

609 See Bosselmann. p. 25 – 29 on the early competition between these concepts.

610 Ibid. p. 26.

sustainability will be implemented and how decisions will be made.[611] As the debate continued, sustainability and sustainable development worked their way to the forefront within the international discourse on environment and development. This debate helped to articulate the current understanding of the concepts and it started in the 1980s.[612]

As mentioned within the development of sustainability and sustainable development, competing agendas appear which complicate integrating economic, social and environmental concerns. It seems that the interpretation of sustainable development depends on the sentiments during a specific era.[613] In the early 1980s different terms emerged as the world was concerned with development and the environment, and an example of such a term was eco-development. Eventually, sustainable development won the day and was popularised by the report of the World Commission on Environment and Development (WCED) in 1987. Before this, however, it became clear that a growing recognition of the ecological consciousness within economic development was needed when the World Conservation Strategy included the following definition of development: 'the modification of the biosphere and the application of human, financial, living and non-living resources to satisfy human needs and improve the quality of human life'.[614]

In 1984, the UN took on the task of developing long-term environmental strategies for the international community. As a result the WCED published its report entitled 'Our Common Future' in 1987, also referred to as the 'Brundtland Report'.[615] Through this report the term sustainable development was popularised.[616] The Brundtland Report is often criticised for not defining sustainability itself but introducing the broader and different concept of sustainable development.[617] The report defined sustainable development as 'development that meets the needs of the present without compromising the ability of future generations to meet their own needs.' The Commission focused its attention on international law and appealed for an overall review of international policy and law founded on sustainable development.[618] As this document had been compiled

611 These two perspectives are often referred to as strong and weak sustainability or the ecologist and environmental approaches.

612 Nico Schrijver, 'Conclusions: The Challenges to Sustainable Development and to Its Further Evolution in International Law', *The Evolution of Sustainable Development in International Law: Inception, Meaning and Status* (2008). p. 211.

613 John Drexhage and Deborah Murphy, 'Sustainable Development: From Brundtland to Rio 2012' (2010). p. 17.

614 World Conservation Strategy: Living Resource Conservation for Sustainable Development, International Union for Conservation of Nature and Natural Resources, United Nations Environment Program, World Wildlife Fund, UNESCO, Food and Agriculture Organization of the United Nations, p. 1, (1980).

615 Brundtland.

616 Schrijver, 'Conclusions: The Challenges to Sustainable Development and to Its Further Evolution in International Law'. p. 208.

617 Bosselmann. p. 25.

618 Christina Voigt, *Sustainable Development as a Principle of International Law* (Martinus Nijhof Publishers 2009). p. 15.

by 21 members from different states, the document aimed to reconcile the North's unsustainable overdevelopment with the South's underdevelopment.[619] For this reason, both ecological sustainability as well as socio-economic justice needed to be reflected in the report. But it seems that sustainable development was purely articulated in terms of human needs, thereby giving it an overly anthropocentric focus and leaving room for trade-offs which adversely affect the environment significantly.[620]

The Brundtland Report was followed by the Rio UN Conference on Environment and Development and it resulted in two soft law instruments which were adopted by 178 member states, the 1992 Rio Declaration on Environment and Development and Agenda 21. These instruments were the first steps in further developing the concepts of sustainability and sustainable development. The Rio Declaration is a clear example of the competing agendas of the Global North and South. On the one hand, the Global South aimed to tackle poverty, trade liberalisation and to increase development assistance, while the Global North focused on climate change and the loss of biodiversity.[621] The Declaration therefore emphasises the connections between the environment and social and economic issues. It formulates 27 principles which aim to guide sustainable development around the globe. Nonetheless, the Declaration lacks definitions of sustainability and sustainable development.[622] It emphasises the integration of environmental and developmental issues by mentioning 'that in order to achieve sustainable development, environmental protection shall constitute an integral part of the development process and cannot be considered in isolation from it.'[623] In addition to this integration principle the Declaration includes the right to development (Principle 3) and poverty alleviation (Principle 5) to name just a few.

Its partner document, Agenda 21, is a global comprehensive action plan for sustainable development and includes 40 chapters covering topics such as conservation and the management of natural resources, the participation of stakeholders and the implementation of the action plan. Through the Agenda the Global North reaffirmed its previous commitment to contribute to official development assistance and to share environmentally sound technologies.[624] Even though the signatories of the Agenda agreed on several principles and concepts, the document lacks a definition of sustainable development. Also, the document was left somewhat open-ended as further commitments where left to further negotiations on institutions and laws. Nevertheless, Agenda 21 has been picked up by local governments as a strategy for implementing sustainable development locally.[625]

The two soft law instruments stemming from the Rio Conference left many questions concerning the definitions and the legal status of sustainability and sustainable

---

619 Bosselmann. p. 29.
620 Ibid. p. 33.
621 Drexhage and Murphy. p. 16.
622 Bosselmann. p. 35.
623 Principle 4.
624 Agenda 21 Chapter 4.
625 Local Agenda 21.

development as neither had any legal effect. Nonetheless, two important legally binding agreements were opened for signature: the UN Framework Convention on Climate Change[626] and the UN Convention on Biodiversity.[627] These two instruments were both landmark environmental agreements embracing a newly developed system for the legal implementation of environmental norms.[628]

The UN Framework Convention on Climate Change does not specifically focus on sustainable development but it does explain that:

> *'The ultimate objective of this Convention and any related legal instruments that the Conference of the Parties may adopt is to achieve, in accordance with the relevant provisions of the Convention, stabilization of greenhouse gas concentrations in the atmosphere at a level that would prevent dangerous anthropogenic interference with the climate system. Such a level should be achieved within a time-frame sufficient to allow ecosystems to adapt naturally to climate change, to ensure that food production is not threatened and to enable economic development to proceed in a sustainable manner.'*[629]

The concept of sustainable development did not only experience developments facilitated by instruments of international policy and international (soft) law but the judiciary also contributed. The most well-known judicial decision concerning sustainable development is that of the International Court of Justice in the 1997 *Gabcikovo-Nagymaros Project* case.[630] The Vienna Convention on the Law of Treaties requires States to uphold standards following from jus cogens. This case was concerned with applying sustainable development to an alleged breach of treaty obligations between Hungary and Slovakia. The Court established that environmental norms are part of the general corpus of international law and the concept of sustainable development can be used to reconcile environmental and developmental conflicts.[631] The Court did not however provide a definition of sustainable development, it merely recognised its utility. Also, the status of sustainable development was not clarified as it was referred to as a concept. The Court's judgment did show that sustainable development should be considered in the decision-making process and can therefore be seen as primarily procedural in nature.[632] Judge Weeramantry added a detailed opinion to this judgment, delving deeper into the topic of sustainable development.

---

626 Which entered into force in 1994.

627 1760 UNTS 79; 31 ILM 818 (1992).

628 Nicholas A Robinson, 'Reflecting on Rio: Environmetal Law in the Coming Decades' in Jamie Benidickson (ed), *Environmental Law and Sustainability after Rio* (Edward Elgar Publishing 2011). p. 10.

629 Article 2.

630 Gabcikovo-Nagymaros (Hungary. v. Slovakia), 1997 available on: www.icj-cij.org/docket/files/92/7375.pdf.

631 Anderson. p. 18.

632 Jessica Howley, 'The Gabcikovo Nagymaros Case: The Influence of the International Court of Justice on the Law of Sustainable Development' (2009) 2 Queensland Law Student Review 1. p. 6.

He characterised sustainable development as a 'principle with normative value'[633] not a mere concept. In addition, he found sustainable development to be instrumental in reconciling where human rights were conflicting, namely: a human right to protection of the environment and a human right to development.[634]

In 2000, all UN member states adopted the Millennium Declaration from which the Millennium Development Goals (MDGs) derive. The MDGs concerned a shared framework for development that is followed by all UN agencies and members. The targets of this framework are geared towards health improvement, gender equality, poverty reduction and environmental protection. One of the goals was to ensure environmental sustainability and specifically in relation to water a target was set to 'halve, by 2015, the proportion of the population without sustainable access to safe drinking water and basic sanitation.'[635] It was said that this target had been reached.[636]

At the 2002 World Summit on Sustainable Development the Johannesburg Plan of Implementation was formulated which set out new commitments in, amongst other areas, poverty reduction, education and institutional arrangements. The Johannesburg Plan authoritatively explained that there is: 'a collective responsibility to advance and strengthen the interdependent and mutually reinforcing pillars of sustainable development – economic development, social development and environmental protection – at local, national, regional and global levels.'[637] The 2002 New Delhi Declaration of Seven Principles of International Law Relating to Sustainable Development is the most current law-oriented benchmark which articulates important principles of the international law of sustainable development. It is based on a comprehensive and balanced study which took place over a decade and was carried out by the International Law Association. In April 2002 the Association presented its findings and stated that sustainable development is a widely accepted global objective which can be found in many international, regional and national provisions. The Declaration includes seven principles[638] which were previously contained in the 1992 Rio Declaration and the Brundlandt Report such as: the principle of common but differentiated responsibilities and the precautionary principle. These principles can be considered to be central to sustainable development. The Johannesburg Plan of Implementation followed the ILA Principles and recognised and reaffirmed the

---

633 Gabcikovo-Nagymaros Project (Hungary v Slovakia) (Separate Opinion of Vice-President Weeramantry) [1997] ICJ Rep. 7, 88 [Separate Opinion]; Separate Opinion,, 88-89, 90.

634 Separate Opinion, 4, 91.

635 Www.un.org/millenniumgoals/environ.shtml.

636 UNICEF/WHO, 2010. Millennium Development Goal drinking water target met. Joint news release, 6 March 2010. Available from: www.who.int/mediacentre/news/releases/2012/ drinking_water_20120306/en/.

637 Www.un-documents.net/jburgdec.htm paragraph 5 Annex.

638 The principles are: the duty to ensure sustainable use of natural resources, principles of equity and the eradication of poverty, the principle of common but differentiated responsibilities, the principle of the precautionary approach to human health, natural resources and ecosystems, the principle of public participation and access to information and justice, the principle of good governance, the principle of integration and interrelationship, in particular in relation to human rights and social, economic and environmental objectives. Paragraph 4.5 will explore these principles.

principles mentioned.[639] This Plan of Implementation is also significant as it also incorporated the concept of the three pillars of sustainable development: economic, environmental and social development.[640]

The 2012 Rio+20, United Nations Conference on Sustainable Development was intended to set a global sustainability agenda for the coming decade. The resulting document of this conference, 'The Future We Want', is considered to fall short of the objectives that were set.[641] The sustainable development framework was in need of a structure, coordination, enforcement and non-state actor participation but the conference failed to deliver on this.[642] And the measures in the resulting document remained of a voluntary nature and reaffirmed previous commitments made in the Rio Principles and previous action plans.[643]

The most recent guiding instrument regarding sustainable development is the 2015 Sustainable Development Goals[644], which set out 17 goals to end poverty, protect the planet and ensure prosperity. The goals and their corresponding targets are aimed to be reached within 15 years. Governments, the private sector, civil society and individuals are supposed to work in unison to make this a reality. The SDGs do not provide a clear definition of sustainability but they however reaffirm the notion that economic development, social development and environmental protection are supportive elements in development.

Economic growth is for a large part dependent on natural resources which are finite and should therefore be managed sustainably. This idea of sustainability was originally focused on the environment. However, other dimensions of human development and welfare were later included. The above-mentioned demonstrates that as a result of these changes, the concept of sustainable development emerged from the idea of sustainability. It brought together the idea that economic growth, social welfare and environmental protection can and should be balanced and reconciled. The concept of sustainable development is still expanding in international law which is seriously required considering the challenges that lie ahead. For instance, the legal instruments available within international law are still inadequate for realising the Johannesburg

639 Official Documents from the World Summit for Sustainable Development; WSSD Documents; World Summit for Sustainable Development (WSSD): Mumbai, Maharashtra, India, August 2002; Available online: www.un.org/jsummit/html/documents/summit_docs.html (accessed on 28 December 2014); see also Johannesburg Declaration on Sustainable Development and Johannesburg Plan of Implementation; Report of the World Summit on Sustainable Development; UN: New York, NY, USA, 4 September 2002; Doc A/CONF.199/L20.

640 Jamie Benidickson and others, *Environmental Law and Sustainability after Rio* (Jamie Benidickson and others eds, Edward Elgar Publishing 2011). p. 2.

641 Https://sustainabledevelopment.un.org/content/documents/733FutureWeWant.pdf.

642 Kailyn Ellison, 'Rio+20: How the Tension Between Developing and Developed Countries Influenced Sustainable Development Efforts' (2014) 27 Global Business & Development Law Journal 107. p. 118. See Clause 19 of The Future We Want.

643 The Future We Want par. 14 – 18.

644 UNGA'Transforming our world: the 2030 Agenda for Sustainable Development'UNDoc A/ RES/70/1 (21 October 2015).

Plan of Implementation.[645] Also, the national application of global principles needs to be further elaborated.[646] Nevertheless, at the moment it is understood that the Brundtland Report contains the most commonly accepted and cited definition of sustainable development and that the requirements which operationalise sustainable development are explained in Agenda 21 and the Rio Declaration but also in subsequent international, regional and national instruments and judgements.

Even though sustainable development has experienced several developments there are still a few major concerns about its articulation. These concerns are often referred to as objections with the first one criticising the flexibility of the concept, the result being that it means different things to different groups (the indeterminacy objection). The ecocentric objection questions whether sustainable development allows effective interventions through environmental law and policy for protecting the environment. Economic growth has been recognised as a major motivator behind sustainable development. Its pursuit, however, has been made subject to environmental conservation and the sustainable use of natural recourses.[647] One of the challenges which sustainability embodies is the balancing of the three elements: social, environmental and economic. Economic development often has priority over the other two elements. However, economic development which degrades its natural resources can only be short-term development. The degradation of natural resources, in this case fresh water, threatens the livelihood of especially vulnerable individuals. And the last objection, the subordination objection, entails the idea that the aspects of sustainable development will be dealt with in a hierarchical manner instead of on an equal playing field.[648]

4.3 LEGAL STATUS AND ENFORCEMENT OF SUSTAINABLE DEVELOPMENT

The legal status of sustainable development and its normative status is a topic of international debate. Furthermore, the way in which sustainable development is to be operationalised is also questioned. There are several leading opinions in the literature. Some authors see sustainable development as being too vague a concept to have a clear

---

645 Nico Schrijver, *The Evolution of Sustainable Development in International Law: Inception, Meaning and Status* (Martinus Nijhof Publishers 2008). p. 221.

646 Maja Goepel, 'Formulating Future Just Policies: Applying the Delhi Sustainable Development Law Principles' (2010) 2 Sustainability 1694 <www.mdpi.com/2071-1050/2/6/1694/> accessed 12 February 2014. p. 1696.

647 Schrijver, 'Conclusions: The Challenges to Sustainable Development and to Its Further Evolution in International Law'. p. 211.

648 Dire Tladi, *Sustainable Development in International Law: An Analysis of Key Enviro-Economic Instruments* (2007). p. 240.
The human rights framework could influence how the aspects of sustainable development can be balanced. Chapter 5 will further explore this.

normative status.[649] While others, such as Sands, argue that sustainable development has become part of international law as a principle of customary law, but how it is applied in practical situations is still to be elaborated.[650] He explains that 'there can be little doubt that the concept of ' sustainable development' has entered the corpus of international customary law, requiring different streams of international law to be treated in an integrated manner.'[651] Sands also argues that sustainable development has both a legal function as well as a procedural one.[652] This analysis is not followed by others who find that sustainable development lacks normative content and therefore a norm-creating character.[653] However, sustainable development can be seen as part of the judicial reasoning process.[654] In this sense it provides a normative concept which can be used when primary norms overlap or conflict and it clarifies how norms should be balanced in a specific context.[655]

What is crucial for determining the legal status of sustainable development and its content is the status of the instruments in which it is formulated. The previous sections show that most of the conceptualisations of sustainable development had been developed through soft-law instruments.[656] As a result some question its applicability and enforceability. Even though soft-law instruments cannot technically be enforced through legal mechanisms, they still have significance and may articulate obligations which are respected.[657] States may even find that following non-legal commitments might be easier to accept.[658] The influence of soft-law instruments should not be underestimated as the inclusion of principles and norms in these instruments can influence customary law.

---

649 A.B.M. Marong, From Rio to Johannesburg: Reflections on the Role of International Legal Norms in Sustainable Development, 21 Geo. Int' L Envrtl. L. Rev. (2003) 57 [hereinafter Marong]. See also Sands, Principles, *supra* note 28, p. 255. See also Birnie & Boyle, *supra* note 19, p. 122.

650 Sands, International Environmental Law, *supra* note 17, pp. 254–255.
Lowe, Unsustainable Arguments, *supra* note 27, p. 23.

651 Sands, International Environmental Law, *supra* note 17, pp. 254.

652 Sands (2012) International Environmental Law p. 208.

653 Lowe, Unsustainable Arguments, *supra* note 27, pp. 24–25.

654 " Norms may function primarily as rules for decision, of concern to judicial tribunals, rather than as rules of conduct. (...) It is in the area of these norms that I believe the search for the normative force of the concept of sustainable development should be sought. Sustainable development can properly claim a normative status as an element of the process of judicial reasoning. It is a meta-principle, acting upon other legal rules and principles--a legal concept exercising a kind of interstitial normativity, pushing and pulling the boundaries of true primary norms when they threaten to overlap or conflict with each other." 1 Lowe, Unsustainable Arguments, *supra* note 27, pp. 24–25. 52.

655 Lowe refers to this as 'interstitial norms'. See Lowe, Politics of Law-Making, in Byers (ed.), The Role of Law in International Politics (2000), pp. 213, 216 and.

656 Dernbach and Mintz. p. 539.

657 Francesca Klug, *A Magna Carta for All Humanity*, vol 226 (Routledge 2015). p. 48.

658 Bothe (1990) " A non-legal commitment is (...) often much easier for a state to accept than a legal one. In all probability, here lies the reason why states do not reject resolutions the terms of which they would by no means accept as a treaty. This presents both an opportunity and a danger. As resolutions also give rise to expectations, they trigger a certain pressure for compliance that is often, as has been shown, effective in the long run. They influence practice, and practice influences law.

The international environmental instruments which have been mentioned do not include measures for enforcement such as sanctions and/or other measures which motivate compliance. 'Modern environmental law assumes that citizens should have standing to challenge the non-enforcement of the law and it assumes that courts have a legitimate role to play in environmental disputes.'[659] This is for instance stimulated by the Aarhus Convention.[660] The enforcement of sustainable development can also take place at the national law level by principles which are translated into national legislation. In this manner sustainable development becomes accessible to individuals at the local level and does not remain a requirement only between states. An increasing number of treaties are recognising the principle of sustainable development and its principles are increasingly being invoked before national courts and international tribunals.[661]

4.4 INTERNATIONAL NORMS INFLUENCING SUSTAINABILITY RELATING TO NATIONAL FRESHWATER RESOURCES

The previous sections have discussed the conceptualisations of sustainable development in general. What is relevant in relation to the realisation of the human right to water is how sustainable development can be applied to water resources. There are several legally binding and non-legally binding instruments which can influence the (sustainable) management of water resources. For this study it is necessary to analyse how the general concept of sustainable development has been conceptualised in relation to water resources. This will assist in identifying the obligations of states in protecting and managing water resources which will support the realisation of the human right to water for vulnerable groups. The following table provides an overview of the relevant instruments. What should be kept in mind is the status of the instruments and whether they are legally binding or not.[662]

---

659 A Dan Tarlock (ed), *Environmental Laws and Their Enforcement* (EOLSS Publishers Co Ltd 2009). p. 56.

660 Convention on Access to Information, Public Participation in Decision-Making and Access to Justice in Environmental Matters, done at Aarhus, Denmark, on 25 June 1998. Available on www.unece.org/fileadmin/DAM/env/pp/documents/cep43e.pdf.

661 An innovative domestic court decision on intergenerational equity, which is a component of sustainable development, is the Philippine Supreme Court Case of Minors Oposa v. Secretary of the Department of Environment and Natural Resources (1993), 33 I.L.M. 173 (1994), [hereinafter Minors Oposa Case]. The case addressed intergenerational equity in the context of the state management of public forest land. See further Bulankulama v. The Secretary, Ministry of Industrial Development (2000), 7 (2) South Asian Envrtl. L. Rep. 1 (Sri Lankan Supreme Court) and Shehla Zia and others v. WAPDA (1992), Case No. 15-K (Pakistan Supreme Court).

662 * indicates which instruments are legally binding.

Table 4.1 Overview relevant instruments shaping sustainable development

Instrument	Provisions
*1971 Convention on Wetlands of International Importance especially as Waterfowl Habitat (Ramsar)**	'fundamental ecological functions of wetlands as regulators of water regimes and as habitats supporting a characteristic flora and fauna, especially waterfowl.'[663] Article 4 Parties to this Convention are obligated to assign suitable water within their territory as wetlands of international importance which will then be included on the list Recommendation 3.3 wise use of wetlands: 'sustainable use of wetlands for the benefit of mankind in a way that is compatible with maintaining the natural properties of the ecosystem'
1972 Declaration of the United Nations Conference on the Human Environment (Stockholm Declaration)	Principle 2: The natural resources of the earth, including the air, water, land, flora and fauna and especially representative samples of natural ecosystems, must be safeguarded for the benefit of present and future generations through careful planning or management, as appropriate. Principle 3: The capacity of the earth to produce vital renewable resources must be maintained and, wherever practicable, restored or improved. Principle 5: The non-renewable resources of the earth must be employed in such a way as to guard against the danger of their future exhaustion and to ensure that benefits from such employment are shared by all mankind.
1992 Rio Declaration on Environment and Development	Principle 2: States have, in accordance with the Charter of the United Nations and the principles of international law, the sovereign right to exploit their own resources pursuant to their own environmental and developmental policies, and the responsibility to ensure that activities within their jurisdiction or control do not cause damage to the environment of other States or of areas beyond the limits of national jurisdiction.
*1992 Convention on Biological Diversity**	Article 6(b): Integrate, as far as possible and as appropriate, the conservation and sustainable use of biological diversity into relevant sectoral or cross-sectoral plans, programmes and policies. Article 8 (d) 'Promote the protection of ecosystems, natural habitats and the maintenance of viable populations of species in natural surroundings';
*1992 UN Framework Convention on Climate Change**	Also, Article 4(e) specifically mentions commitments in relation to water resources when it encourages: to 'develop and elaborate appropriate and integrated plans for coastal zone management, water resources and agriculture, and for the protection and rehabilitation of areas, particularly in Africa, affected by drought and desertification, as well as floods.
*1992 UNECE Convention on the Protection and Use of Transboundary Watercourses and International Lakes**	Article 3: To prevent, control and reduce transboundary impact, the Parties shall develop, adopt, implement and, as far as possible, render compatible relevant legal, administrative, economic, financial and technical measures, in order to ensure, inter alia, that: (b) Transboundary waters are protected against pollution from point sources through the prior licensing of waste-water discharges by the competent national authorities, and that the authorised discharges are monitored and controlled; (i) Sustainable water-resources management, including the application of the ecosystems approach, is promoted;

[663] Preambulatory clause.

Instrument	Provisions
1992 Agenda 21	Chapter 18 Protection of the Quality and Supply of Freshwater Resources: Application of Integrated Approaches to the Development, Management and Use of Water Resources
*1994 United Nations Convention to Combat Desertification in Those Countries Experiencing Serious Drought and/ or Desertification, Particularly in Africa**	Article 4 The Parties shall implement their obligations under this Convention, individually or jointly, either through existing or prospective bilateral and multilateral arrangements or a combination thereof, as appropriate, emphasising the need to coordinate efforts and develop a coherent long-term strategy at all levels.
*1997 UN Convention on the Law of the Non-navigational Uses of International Watercourses**	Watercourse States shall in their respective territories utilise an international watercourse in an equitable and reasonable manner. In particular, an international watercourse shall be used and developed by watercourse States with a view to attaining optimal and sustainable utilisation thereof and benefits therefrom, taking into account the interests of the watercourse States concerned, consistent with adequate protection of the watercourse.
*1999 Protocol on Water and Health to the 1992 Convention on the Protection and Use of Transboundary Watercourses and International Lakes**	Article 5 (d) Water resources shall be managed so that the needs of the present generation are met without compromising the ability of future generations to meet their own needs; Article 5(e): Protect water resources used as sources of drinking-water, and their related water ecosystems, from pollution;
IUCN 2010 Draft International Covenant on Environment and Development[664]	'Parties shall take all appropriate measures to maintain and restore the quality of all forms of water, including both salt and fresh water, whether contained in the atmosphere, the oceans, in underground aquifers or watercourses such as lakes and rivers to meet basic human needs and as an essential component of aquatic systems. Parties also shall take all appropriate measures, in particular through integrated conservation and management of water resources and appropriate sanitary measures, to ensure the availability of sufficient quantities of water to satisfy basic human needs and to maintain aquatic systems.'
The 2015 Sustainable Development Goals	Targets of goal no. 6 – Improve water quality by reducing pollution, eliminating dumping and minimising release of hazardous chemicals and materials, halving the proportion of untreated wastewater and substantially increasing recycling and safe reuse globally – Substantially increase water-use efficiency across all sectors and ensure sustainable withdrawals and supply of freshwater to address water scarcity and substantially reduce the number of people suffering from water scarcity – Implement integrated water resources management at all levels, including through transboundary cooperation as appropriate – Protect and restore water-related ecosystems, including mountains, forests, wetlands, rivers, aquifers and lakes – Support and strengthen the participation of local communities in improving water and sanitation management

664 The first edition was published in 1995.

Even though some of these provisions and the instruments which they are part of are not legally binding, they do however provide guidance on how states are meant to exercise their sovereign rights over natural resources. These norms may not currently be strict obligations. Nevertheless their inclusion in instruments like these in the future may assist them in evolving into legally binding norms, for instance by way of customary law.[665] These non-legally binding norms include potential duties ranging from safeguarding water as a natural resource for the benefit of present and future generations to taking measures to protect the recharge areas for groundwater and maintaining the ecological function of forests to control erosion and floods.[666]

The legally binding instruments included more specific legal obligations. However, these obligations depend on the specific sets of circumstances and the objective in question. For instance, some instruments have specifically stated that they apply to shared water courses or that they have the purpose of protecting public health. What can therefore be concluded is that national water resources are managed in accordance with norms and rules relating to the environment and the development of natural resources and not by a particular set of norms that apply specifically to water resources.[667]

4.5 PRINCIPLES OF SUSTAINABLE DEVELOPMENT

Now that the history of sustainable development in international law and policy has been discussed, the specific environmental principles that have emerged in order to achieve sustainability will be shortly looked at. There is a wealth of literature on the separate principles and sustainable development in general. The following sections, will however, discuss these in a concise manner which will be sufficient to connect the principles to the human rights framework in the next chapter.

Even though the Brundtland Report provides a definition, a further delineation is needed for states to modify their actions and meet their obligations. Several principles of environmental law give substance to sustainable development. The International Law Association contributed to the further articulation by identifying general substantive and procedural principles of international law for sustainable development.

What should be kept in mind is that some principles are more relevant than others in the pursuit of sustainable water management. For this study it is relevant to consider an interpretation of the principles which can be translated into the water context, how these principles can be used in the legal context and which dilemmas it might create in relation to the human right to water. The following section takes the first step and identifies the relevant environmental principles for water resources management. In the

665 Owen Mcintyre, 'The Role of Customary Rules and Principles of International Environmental Law in the Protection of Shared International Freshwater Resources' (2006) 46 Natural Resources Journal 158. p. 162.

666 See Douglas Fisher, *The Law and Governance of Water Resources* (Edward Elgar Publishing 2009). p. 112 for an overview of non-legal norms.

667 Ibid. p. 114.

next chapter it will be discussed how these principles can be applied in order to facilitate sustainable access to water for individuals and vulnerable groups and what could be the possible challenges.

There are 8 main principles[668] which can be identified as part of sustainable development:

1. the duty of States to ensure sustainable use of natural resources;
2. equity and the eradication of poverty;
3. common but differentiated responsibilities;
4. precautionary approach to human health, natural resources, and ecosystems;
5. public participation and access to information and justice;
6. good governance
7. integration and interrelationship, particularly in relation to human rights and social, economic, and environmental objectives.

In addition to these principle, the polluter pays principle has been added as the 8th principle because of its specific relevance for water resources.

As mentioned, sustainable development has been defined as: 'development that meets the needs of the present without compromising the ability of future generations to meet their own needs'.[669] This definition of sustainable development therefore includes three notions. Firstly, that the needs of the present population must be met and that special attention should be given to the poor.[670] Secondly, that the current situation has inherent limitations for the ability of the environment to meet the needs of future and present populations. And lastly, the needs of the present and future generations need to be balanced.[671] These aspects have been further conceptualised into the principles of intragenerational equity, integration and intergenerational equity. In order to bring these aspects to life several principles have been formulated to assist the process which will be discussed in the following sections.

1. The duty of States to ensure the sustainable use of natural resources

As explained, the foundation of sustainable development lies in the sustainable use of natural resources such as forests and fish.[672] As a result of the report 'Limits of Growth', 'the concept of optimal and rational use of natural resources in a more general sense became popular and, moreover, was applied to all natural resources, living and non-

668 Other authors like Schrijver, 'Conclusions: The Challenges to Sustainable Development and to Its Further Evolution in International Law'. p. 208 use different categories.

669 Brundtland Report.

670 See Steve Rayner and Elizabeth L Malone, 'Climate Change, Poverty, and Intragenerational Equity: The National Level' (2001) 1 International Journal of Global Environmental Issues 175.

671 Duwell. p. 551.

672 Schrijver, 'Conclusions: The Challenges to Sustainable Development and to Its Further Evolution in International Law'. p. 209.

living and renewable and non-renewable.'[673] Even though states have the sovereign right to manage their natural resources in whichever way they see fit.[674] There is still a legal responsibility for these states to ensure that they use natural resources sustainably and for the development of their citizens.[675] A central concern of this principle is the regenerative capacity of natural resources.

2. *Equity and the eradication of poverty*

The early conceptualisation of sustainable development in the Brundtland Report contains the central principle of intergenerational equity.[676] Intergenerational equity[677] is a concept which has been used within different contexts and is known throughout different nations. In the sustainability and environmental context, intergenerational equity refers to the fairness in the inter-temporal distribution of natural wealth, and its burdens, obligations and rights in relation to its exploitation.[678]

Intergenerational equity, as an emerging principle, is part of international law and is considered to be an essential requirement of sustainable development. Intergenerational equity has been included in several international instruments such as the Stockholm Declaration and the Rio Declaration.[679] From a legal perspective, one of the major challenges within the concept of intergenerational equity are the obligations and responsibilities towards future generations.[680] Nevertheless, the recognition of intergenerational equity has been formally accepted and has nationally acquired the force of law in several instances.[681] This concept has been included in international treaties and has been applied in court cases. Several states have also committed

---

673 Ibid. p. 209.

674 (Onzivu, 2012) p. 877.

675 International Law Association, 'ILA New Delhi Declaration of Principles of International Law Relating to Sustainable Development' (2002) 2 International Environmental Agreements: Politics, Law and Economics 211. p. 213.

676 One of the earliest references is the Seventh Generation philosophy of the Iroquois Confederations, obligating members of this Native nation to consider the impact of every decision made on the seventh generation to come Shelton H Davis, 'Indigenous Peoples and Climate Change' (2010) 1 The International Indigenous Policy Journal 1. p. 8.

677 For literature on intergenerational equity see: Dinah Shelton, 'Intergenerational Equity' in Rudiger Wolfrum and Chie Kojima (eds), *Solidarity: a Structural Principle of International Law* (Springer Berlin Heidelberg 2010).; Edith Brown Weiss, 'The Planertary Trust: Conversation and Intergenerational Equity' (1984) 11 Ecology Law Quarterly 495.; Zena Hadjiargyrou, 'A Conceptual and Practical Evaluation of Intergenerational Equity in International Environmental Law' (2016) 18 International community law review 248.

678 Edith Brown Weiss, 'Intergenerational Equity : A Legal Framework for Global Environmental Change' in Edith Brown Weiss (ed), *Environmental change and international law: New challenges and dimensions* (United Nations University Press 1992). p.6.

679 Emilio Padilla, 'Intergenerational Equity and Sustainability' (2002) 41 Ecological Economics 69. p. 73.

680 See Duwell. l.

681 Weiss, 'Intergenerational Equity : A Legal Framework for Global Environmental Change'. p. 4.

themselves to protecting the interest of future generations by including such obligations in higher and lower laws and policies.[682]

In the case of courts, several influential tribunals have recognised intergenerational equity as an emerging principle of international law when it concerns natural resources. The opinions of Judge Weeramantry of the International Court of Justice concerning maritime boundaries and nuclear testing have often been referred to.[683] Even though the concept did not significantly contribute as a decisive factor, its mentioning still indicated its possible direct application in future decisions concerning natural resources. But national courts are also contributing. For instance, the Philippines Supreme Court referred to the rights of future generations to a healthy environment in a case where the government had issued timber licences.[684] That case was a class-action suit brought also on behalf of future generations.

Nevertheless, unlike determining the needs of the present generation, it is challenging to determine the needs of future generations. It is likely that their needs will differ from those of the present generation; however, history has shown us that without basic environmental services human life is impossible.[685] Services such as clean air, accessible water and fertile soils can be identified as being essential as they make up the services which enable the environment to meet human needs. From this we may conclude that even though we do not know how much water future generations will need or how they will use it, the principle that they will need it cannot be questioned.[686] And our actions or inaction today directly affect their access. Especially since water is a relative finite resource[687] which knows no substitute.

Intra-generational equity[688] is also a principle of sustainable development. Intra-generational equity in the context of sustainable development is concerned with the fair distribution of the outcomes of development within one generation both between people within one national society as well as between nations.[689] From this definition several elements can be distilled, namely: fair distribution relating to the outcome of development[690], addressing social inequity and North-South relations.

---

682 For instance New Zealand has a Parliamentary Commissioner for the Environment which takes the needs of future generations into account.

683 See Gabičkovo-Nagymaros (Hungary/Slovakia) ICJ Reports. Reprinted 37 ILM. 162 (1997), p. 206.

684 'Human Rights and the Environment', *Proceedings of the 13th Informal ASEM Seminar on Human Rights* (Asia-Europe Foundation (ASEF) 2013) <www.pubmedcentral.nih.gov/articlerender.fcgi?artid=3638120&tool=pmcentrez&rendertype=abstract>. p. 64.

685 Bosselmann.p. 31.

686 Idem.

687 Because for instance replenishment of groundwater after depletion can take generations.

688 For literature on intra-generational equity see: GF Maggio, 'Inter/Intra-Generational Equity : Current Applications under International Law for Promoting the Sustainable Development of Natural Resources' 4 Buffalo environmental law journal 161.; Dire Tladi, 'Intragenerational Equity : A New Name for International Environmental Justice' (2003) 9 Freedom, Justice and Equality : three Pillars of Legal History 197.

689 E Feitelson and J Chenoweth, 'Water Poverty: Towards a Meaningful Indicator' (2002) 4 Water Policy 263 <http://linkinghub.elsevier.com/retrieve/pii/S1366701702000296>. p. 268.

690 Related to the principle of public participation is the principle of access and benefit sharing regarding natural resources. Both are mentioned in Principle 10 which was mentioned above. This principle is particularly mentioned in relation to indigenous peoples and local communities. These communities

The concept has so far been mostly approached from an economic point of view, focusing on the imbalance in wealth between North and South and poverty reduction within nations. Poverty reduction, even though not always expressly referred to as an issue of intra-generational equity, is often found in instruments as one of the objectives of sustainable development and as a cause of unsound environmental actions by individuals. The Brundtland Commission explains that poverty forces those suffering from it to act environmentally unsustainably when it states that:

> *'Those who are poor and hungry will often destroy their immediate environment in order to survive: They will cut down forests; their livestock will overgraze grasslands; they will overuse marginal land; and in growing numbers they will crowd into congested cities. The cumulative effect of these changes is so far-reaching as to make poverty itself a major global scourge.'*

Development can only be sustainable when both inter-generational and intra-generational equity are achieved through integration.[691]

3. *Common but differentiated responsibilities*

The principle of common but differentiated responsibilities[692] is an example of solidarity and equity among states. This principle recognises the historical differences in the contributions made by developing and developed nations to global problems and the differences in ability to deal with these problems. The principle thereby contains two elements. Firstly, it emphasises the common responsibility of states in protecting the environment at the national, regional and international level. Secondly, it takes into account the different circumstances and capabilities of states. It considers their contribution to environmental degradation and their ability to prevent, reduce and control the threat.[693]

The recognition of the special situation of developing states first emerged in the 1987 Montreal Protocol.[694] Since then the principle is often mentioned in relation to

have strong ties to these resources and are often aware of interacting with them in a sustainable manner. With the degradation of natural resources their livelihoods and cultures are at risk of disappearing. Similarly, human rights norms set out that indigenous peoples and local communities have the right to participate in decision-making processes.

691 Cosens and others. p. 2346.

692 For literature on the principle of common but differentiated responsibilities see: Thomas Deleuil, 'The Common but Differentiated Responsibilities Principle: Changes in Continuity after the Durban Conference of the Parties' (2012) 21 Review of European Community and International Environmental Law 271; Mary J Bortscheller, 'Equitable But Ineffective: How The Principle Of Common But Differentiated Responsibilities Hobbles The Global Fight Against Climate Change' (2010) 10 Sustainable Development Law & Policy 49 <http://digitalcommons.wcl.american.edu/cgi/viewcontent.cgi?article=1036&context=sdlp>.

693 Sands (1995) p. 334.

694 Article 5 of the Montreal Protocol on Substances that Deplete the Ozone Layer, Sept. 1987 The principle is also part of other fields of international law such as international economic law, see for instance Part IV of the General Agreement on Tariffs and Trade (GATT). But also "has a firm status in various fields of international law, including human rights law, international trade law and

climate change. In 1992, developing states insisted on the principle being included in the UNFCCC as they were concerned about who would initiate reducing their greenhouse gas emissions. Developing states historically did not contribute very much to the global greenhouse stock but more recently depend on activities which produce greenhouse gases and could not therefore commit themselves to high targets. Including this principle in the UNFCCC made it more inclusive and helped it to gain legitimacy.

Principle 7 of the Rio Declaration states:

> *'States shall cooperate in a spirit of global partnership to conserve, protect and restore the health and integrity of the Earth's ecosystem. In view of the different contributions to global environmental degradation, States have common but differentiated responsibilities. The developed countries acknowledge the responsibility that they bear in the international pursuit of sustainable development in view of the pressures their societies place on the global environment and of the technologies and financial resources they command.'*

This formulation also makes clear that the principle of common but differentiated responsibilities does not stand by itself but is seen in the bigger picture of cooperation between states and often demands financial assistance and access to technology.[695] In practical terms, this principle requires states to take on measures aimed at addressing environmental problems (common responsibilities). And these measures will allow for differing obligations and standards for states (differing responsibilities).

4. *The precautionary approach to human health, natural resources, and ecosystems*

The precautionary principle[696] is closely connected to intergenerational equity and is a precondition for sustainable development. The general idea of the precautionary principle is concerned with serious environmental and health threats and it finds that these hazards should be anticipated and prevented before the damage is realised even in the event that scientific knowledge of the risk is inadequate.[697] This is the general idea; however, the precautionary principle takes various forms. Several definitions and formulations of the principle can be found in international environmental law instruments and several definitions have been proposed in academic literature. The 1982 World Charter for Nature adopted by the UN General Assembly included the

---

international environmental law', ILA Committee on Legal Aspects of Sustainable Development (2002), p. 9 16, 1987.

695 ILA Principles paragraph 3.4.

696 For literature on the precautionary principle see: Arie Trouwborst, 'The Precautionary Principle in General International Law: Combating the Babylonian Confusion' (2007) 16 Review of European Community and International Environmental Law 185. Arie Trouwborst, 'Precautionary Rights and Duties of States' (Utrecht University 2006) D. Freestone, 'The Continued Rise of the Precautionary Principle', in A. Boyle and D. Freestone (eds), International Law and Sustainable Development (Oxford University Press, 1999); David Freestone and Ellen Hey (eds), *The Precautionary Principle and International Law : The Challenge of Implementation* (Kluwer Law International 1996).

697 Kiefer and others, *Legal Resources for the Right to Water and Sanitation.*

first international recognition of the precautionary principle. Article 11 (b) states: "Activities which are likely to pose a significant risk to nature shall be preceded by an exhaustive examination; their proponents shall demonstrate that expected benefits outweigh potential damage to nature, and where potential adverse effects are not fully understood, the activities should not proceed."

The principle has since then been reflected in several international instruments. What should first be explained is the difference between the precautionary principle and the precautionary approach. Principle 15 of the Rio Declaration explains that a precautionary approach should be taken in the context of scientific uncertainty and according to the capacity of states. The Declaration goes on to state: 'Where there are threats of serious or irreversible damage, lack of full scientific certainty shall not be used as a reason for postponing cost-effective measures to prevent degradation.[698] The principle is concerned with ensuring a greater allowance for uncertainty in regulating environmental risks and in relation to the use of natural resources. This definition adopts an economic rationale which again shows the unequal relationship between balancing environmental and economic concerns.

Within hard and soft law there are multiple instruments which subscribe to and expand on this notion of precaution. For this reason, some argue that the principle of precaution is an international legal principle enjoying general application.[699] However, there is no general agreed notion of precaution which is applicable in all contexts. But the basic idea of this concept includes the following elements:

1. Scientific uncertainty is not a justifiable excuse for neglecting to take action in relation to environmental concerns
2. Particular environmental concerns require positive action by States.
3. When potentially damaging activities are taking places the burden of proof is on those engaging in these activities to establish the absence of environmental harm.
4. States are allowed to restrict imports in relation to possible significant environmental harm.[700]

Several cases have emerged which assist in showing how this principle is applied and explaining the status of the principle, the threshold requirements and the location of the burden of proof. For instance, in India the Vellore Citizens Welfare Forum filed a public interest case arguing that tanneries and other industries were dumping untreated waste water onto agricultural fields, roads and into waterways. As a result, local drinking water resources became contaminated. It was decided that in such a case

698 Dellapenna and Gupta. p. 14.

699 Dinah Shelton and Alexandre Kiss, *Judicial Handbook on Environmental Law* (UNEP 2005) <www.unep.org/delc/Portals/119/publications/Judicial-Handbook-Environmenal-Law.pdf>. p. 13 Vellore Citizens Welfare Forum v. Union of India, [1996] AIR SC 2715 (finding the principles of sustainable development, the polluter pays and precaution to be part of customary international law).

700 Ibid. p. 21.

the burden of proof is on the developer or industrialist to prove that his action will not cause environmental harm.[701]

This and other cases show how this principle can be used locally. National and local governments could implement this principle through legislation and take it into consideration when awarding concessions for the exploitation of natural resources and other activities which might adversely affect the environment. Individuals or groups could, as was the case in the Vellore decision, bring their concerns under the topic of public interest if the law in the relevant state allows for this.

5. *Public participation and access to information and justice*

Principle 10 of the Rio Declaration states that:

> *'Environmental issues are best handled with the participation of all concerned citizens, at the relevant level. At the national level, each individual shall have appropriate access to information concerning the environment that is held by public authorities, including information on hazardous materials and activities in their communities, and the opportunity to participate in decision-making processes. States shall facilitate and encourage public awareness and participation by making information widely available. Effective access to judicial and administrative proceedings, including redress and remedy, shall be provided.'*

Even though Principle 10 is not binding, it is a good point of reference against which national standards for public participation can be measured, such as national legislation. This notion of public participation[702] has been supported by the commonly referred to 1998 Aarhus Convention on Access to Information, Public Participation in Decision-Making and Access to Justice in Environmental Matters. The Convention was drafted by the United Nations Economic Commission for Europe in 1998. The Aarhus Convention was ratified by the EU in 2005 and turned into law by Regulation 1367/2006. Other states are also parties to the Convention including several Caribbean states.[703]

The Convention provides for several obligations that are applicable in the national jurisdictions of state parties and which are relevant to individuals and groups. Firstly, under Article 9(1) of the Convention access to justice needs to be granted in relation to requests for environmental information. The public should be able to complain if a request for information has not been appropriately dealt with. Secondly, Article 9(2) of

701 Vellore Citizen Welfare Forum vs Union of India & Ors., W.P.(C) No. 914/1991.

702 For literature on public participation and access to information and justice see: Carl E Bruch, *Public Participation in the Governance of International Freshwater Resources* (United Nations University Press 2005); Jona Razzaque, 'Public Participation in Water Governance' in Joseph W Dellapenna and Joyeeta Gupta (eds) (Springer Netherlands 2009); Frans Coenen, 'Local Agenda 21 : " Meaningful and Effective " Participation ?' in Frans Coenen (ed), *Public Participation and Better Environmental Decisions* (Springer Netherlands 2009); Andres Harding, *Access to Environmental Justice: A Comparative Study* (Nijhoff 2007).

703 Anderson. p. 57.

the Convention requires states to grant access to 'a review procedure before a court of law or another independent and impartial body.' Thirdly, Article 9 (3) contains a right of access to the national system. These obligations provide options for individuals to contest actions of individuals and governments. However, the Convention does depend on the implementation of environmental goals at the national level.

The Rio Declaration contributed several other aspects to the contemporary debate on public participation. Firstly, it underlined the link between human rights and environmental concerns and does so by encouraging individuals to use human rights procedures to bring about environmental protection. As a result, international human rights institutions noticed an increase in individuals seeking recourse for environmental issues. Secondly, the Declaration contributed by recognising the special rights and interests of groups such as indigenous communities.[704] Their land rights and traditional knowledge are connected to preserving biodiversity. In several cases this discourse has proved to be useful in protecting the environment.[705]

6. *Good governance*

Good governance[706] is essential for attaining sustainability goals. Like sustainable development, good governance has not been strictly formulated. Several instruments have identified aspects which are relevant to good governance. The ILA Principles state that transparent[707] and democratic decision-making processes should be established. In addition, effective accountability is also very important. Corruption, a lack of accountability and limited financial resources are often identified as undermining effective public participation and hampering reform in the water sector.[708] For example, in certain cases investments are made but due to unclear departmental responsibilities they do not attain their objective which undermines state commitments to pursuing sustainable development.

7. *Integration and its interrelationship, particularly in relation to human rights and social, economic, and environmental objectives*

The core of sustainable development is understanding that environmental and developmental concerns are inextricably interrelated. It requires that both

---

704 See Rhett B Larson, 'Water, Worship, and Wisdom: Indigenous Traditional Ecological Knowledge and the Human Right to Water' (2012) 19 ILSA Journal of International and Comparative Law 43. on the knowledge within indigenous communities.

705 Jeremy Sarkin and Amelia Cook, 'The Human Rights of the San (Bushmen) of Botswana – the Clash of the Rights of Indigenous Communities and Their Access to Water with the Rights of the State to Environmental Conservation and Mineral Resource Exploitation' (2011) 20 Journal of International Law and Policy 1.

706 For literature on good governance see: OECD, 'OECD Principles on Water Governance' (2015); Zhang Wei (ed), *Human Rights and Good Governance* (Brill Nijhoff 2016).

707 See Anoeska Wilhelmina Geertruida Johanna Buijze, 'Thee Principle of Transparency in EU Law' (Utrecht University 2013) <http://dspace.library.uu.nl/handle/1874/269787>. See also Chapter 2.

708 Onzivu. p. 886.

environmental and development considerations are taken into account in the decision-making process resulting in an holistic approach. The Stockholm Declaration has several principles which mention integration. For instance, Principle 10 refers to ecological processes and economic factors to be taken into account by developing states, thereby recognising the necessity of integration.[709]

The Rio Declaration states that: 'in order to achieve sustainable development, environmental protection shall constitute an integral part of the development process and cannot be considered in isolation from it.'[710] The Declaration notably saw a role for judicial instruments that will influence the environment and economic development planning. Agenda 21 also calls for the 'integrated enforcement of effective laws and regulations.'[711]

Integration and its interrelationship within sustainable development can provide for several challenges and points of criticism. Firstly, finding a balance between the three pillars is frequently seen as unrealistic. Sustainable development is often criticised for allowing preferential treatment for economic growth over the pillars. An example supporting this criticism can be found in Principle 12 of the Rio Declaration which reads: 'Trade policy measures for environmental purposes should not constitute a means of arbitrary or unjustifiable discrimination or a disguised restriction on international trade.' This article sets out that trade and the connected economy are not to be compromised by environmental efforts. However, the Rio Declaration does not mention a restriction on economic targets because of environmental degradation. Others argue that in order for sustainable development to be conceptualised there needs to be an emphasis on one of the pillars.

Secondly, integration and its interrelationship imply a consideration of all actions within a state which contribute to economic growth and may influence the environment. These actions are often coordinated or witnessed by different parties, at different levels. As a result national and local governments are often confronted with institutional difficulties. These difficulties are caused by the multiplicity of administrative responsibilities resulting in overlaps, gaps and internal conflicts over who will take on which responsibilities.[712] In the event that laws and regulations do not cause fragmentation it is still necessary that coordination takes place to ensure that the implementation of laws is consistent with the goals for sustainable development that states have committed themselves to.[713]

---

709 Principle 10: For the developing countries, stability of prices and adequate earnings for primary commodities and raw materials are essential to environmental management, since economic factors as well as ecological processes must be taken into account.
Also see Principles 8 and 13.

710 Principle 4.

711 Paragraph 8.14.

712 Cecilia Tortajada, 'IWRM Revisited: From Concept to Implementation' (2014) 30 International Journal of Water Resources Development 361 <www.tandfonline.com/doi/abs/10.1080/07900627.2014.937085> accessed 28 August 2014. p. 362.

713 Anderson. p. 50.

8. *The Polluter Pays Principle (PPP) and the Natural Resource User Pays*

Another principle which is only mentioned once in the ILA Principles, in relation to the principle of common but differentiated responsibilities, but is nevertheless highly relevant to the sustainable management of water is the Polluter Pays Principle.[714] This Principle and its variation, the more recent concept of the Natural Resource User Pays, take an economic approach to preventing pollution and the over-exploitation of natural resources.[715] The principles have been expressly recognised in various multilateral conventions, both in the preambles as well as in the operative clauses.[716]

Article 16 of the Rio Declaration explains that:

> *'national authorities should endeavour to promote the internalization of environmental costs and the use of economic instruments, taking into account the approach that the polluter should, in principle, bear the cost of pollution, with due regard to the public interest and without distorting international trade and investment.*[717]

The User Pays Principles further develops this principle and calls upon the user to bear the costs of exploiting natural capital.[718] The Polluter Pays Principle is currently not a recognised principle of international law. Nevertheless, this principle can be a leading factor in the development of policies.[719] The reasoning behind these principles is that the user should be held responsible for its actions by internalising environmental and social costs.[720] In other words, those who are degrading the environment are to compensate for it as well.[721]

---

714 For literature on the Polluter Pays Principle see: Petra E Lindhout and Berthy van Den Broek, 'The Polluter Pays Principle: Guidelines for Cost Recovery and Burden Sharing in the Case Law of the European Court of Justice' (2014) 10 Utrecht Law Review 46; Priscilla Schwartz, 'Principle 16: The Polluter-Pays Principle' in Jorge E Vinuales (ed), *The Rio Declaration on Environment and Development : a Commentary* (Oxford University Press 2015); Mcintyre. p. 197; Andri Wibisana, 'Three Principles of Environmental Law : The Polluter-Pays Principle, the Principle of Prevention, and the Precautionary Principle', *Environmental Law in Development: Lessons from the Indonesian Experience* (Elgar Publishers 2006).

715 Agenda 21. par. 8.28 'During the past several years, many Governments, primarily in industrialized countries but also in Central and Eastern Europe and in developing countries, have been making increasing use of economic approaches, including those that are market -oriented. Examples include the polluter-pays principle and the more recent natural-resource-user-pays concept.'

716 De Sadeleer (2002),p. 23.

717 Principle 16 Rio Declaration.

718 Glossary of Environment Statistics, Studies in Methods, Series F, No. 67, United Nations, New York, 1997.

719 Antionette Hildering, *International Law, Sustainable Development and Water Management* (1st edn, Eburon Uitgeverij BV 2004). p. 159.

720 This principle is often connected to Principle 8 Rio Declaration which wants to 'eliminate unsustainable patterns of production and consumption' as the internalization of costs can lead to this.

721 Stephen J Turner, *A Global Environmental Right* (Routledge 2014). p. 77.

4.6 CONCLUSION

Sustainable development has developed out of the more general norm of sustainability. It has come a long way since its original meaning of the sustainable use of natural resources and has moved towards a more anthropocentric and socio-economic focus.[722] When it comes to the use of the terms sustainable development and sustainability, this study takes the perspective that sustainability can be seen as the long-term end goal while sustainable development has shaped the processes which need to take place in order to attain sustainability.[723] This study therefore uses sustainable development as articulated in the Brundtland Report. The legal implications of sustainable development have not yet been fully defined. However, seven principles which operationalise the otherwise vague concept have been identified: the duty of States to ensure the sustainable use of natural resources; equity and the eradication of poverty; common but differentiated obligations; the precautionary approach to human health, natural resources, and ecosystems; public participation and access to information and justice; good governance; and integration and its interrelationship. These principles will be used to inform the sustainable realisation of the human right to water for vulnerable groups.

722 Schrijver, 'Conclusions: The Challenges to Sustainable Development and to Its Further Evolution in International Law'. p. 217.

723 It has been said that sustainable development is the pathway to sustainable development.

CHAPTER 5

ENHANCING THE HUMAN RIGHT TO WATER

Finding Inspiration from the Concept of Sustainable Development

'The manner of the realization of the right to water must also be sustainable, ensuring that the right can be realized for present and future generations.'[724]

5.1 INTRODUCTION

The previous chapters have set the stage for answering the (main) research question: how can a sustainable human right to water be formulated for the enjoyment of vulnerable groups? This study finds the answer by supplementing the human right to water with elements from the concept of sustainable development. Elements discussed in previous chapters can support the implementation and further development of the human right to water for vulnerable groups. Chapter 2 explains which individuals can be considered vulnerable and focusses especially on those who are currently unserved or underserved. These individuals are not only at risk of having their human right to water infringed due to financial or social constraints. Vulnerable groups experience an additional risk when they are unable to depend on their natural surroundings to meet their water needs due to, for instance, drought or pollution. For this reason, connecting environmental vulnerability to social vulnerability can provide a more effective approach in safeguarding the human right to water for all individuals and groups at risk.

Chapter 3 provided a detailed overview of the content of the human right to water and states' obligations. It was also explained that access to water originated from environmental law and policy and developed into a human right. However, along the way it has lost elements which take environmental factors into account. These environmental factors are however highly relevant for the effective realisation of the human right to water. Taking environmental factors into account will allow the right to be effectively implemented in the natural context where it is supposed to function.

---

724 Committee on Economic Social and Cultural Rights.

For this reason, this study seeks inspiration from the concept of sustainable development which was explained in Chapter 4. Here the content and its connected principles were analysed. The current chapter builds on the previous chapters and argues for the inclusion of elements of sustainable development and its alignment with human rights. These approaches add to the human right to water as they can be used to facilitate sustainable water management which is necessary for the human right to water to function in today's water-scarce context. This chapter therefore focusses on how sustainable development can assist in implementing and realising the human right to water. It discusses the possible compatibilities and conflicts between the two approaches: the human rights approach and sustainable development.

Another way in which different approaches relevant to water can be integrated is through Integrated Water Resources Management (IWRM).[725] This entails 'a process which promotes the coordinated development and management of water, land and related resources, in order to maximise the resultant economic and social welfare in an equitable manner without compromising the sustainability of vital ecosystems.'[726] IWRM has been accepted as an approach to safeguarding 'economically sound and environmentally sustainable management of water resources and provision of water services.' Many argue that IWRM should be the dominant approach in sustainable water use management.[727] However, there are several challenges in adopting this approach. It has not yet been fully articulated and questions remain as to its practical implementation.[728] For this study another approach has been chosen, namely encompassing certain elements from sustainable development, as has been explained. This is because General Comment No. 15 refers to the *sustainable* realisation of the human right to water.[729] In addition, the former Special Rapporteur has expressed concern about IWRM paying insufficient attention to its social equity dimension.[730]

---

725 IWRM is guided by the four Dublin Principles in 1992:
1. Fresh water is a finite and vulnerable resource, essential to sustain life, development and the environment.
2. Water development and management should be based on a participatory approach, involving users, planners and policymakers at all levels.
3. Women play a central part in the provision, management and safeguarding of water.
4. Water has an economic value in all its competing uses and should be recognized as an economic good.

726 Global Water Partnership (2000), Integrated Water Resources Management. Stockholm: Global Water Partnership p. 22.

727 See for instance AK Biswas, 'Integrated Water Resources Management: Is It Working?' (2008) 24 International Journal of Water Resources Development p. 7 IWRM is also mentioned as a central approach in the Sustainable Development Goals UN, 'The Sustainable Development Goals Report 2016' (2016) <www.theguardian.com/global-development/ng-interactive/2015/jan/19/sustainable-development-goals-changing-world-17-steps-interactive>.

728 AK Biswas, 'Integrated Water Resources Management: Is It Working?' (2008) 24 International Journal of Water Resources Development p. 8.

729 Committee on Economic Social and Cultural Rights. paragraph 7.

730 De Albuquerque and Roaf, *On The Right Track. Good Practices in Realising the Rights to Water and Sanitation*. p. 118.

Also, this study concentrates on water for domestic purposes as a central issue and thus a particular element of water management.

Figure 5.1 Representation of the distant obligations and core of the human right to water

Ensuring vulnerable groups' sustainable access to water for domestic purposes means that several aspects in relation to water resources management must be taken into account. This chapter discusses the added value of sustainable development for the realisation of the human right to water. The concept of sustainable development supports the human right to water by helping to define the outer boundaries of its substantive scope, for instance obligations which have implications for water management. These obligations which stem from sustainable development can be referred to as the '*distant corollaries*' of the right to water.[731] We can also refer to obligations that are an obvious but remote consequence of respecting, protecting, and fulfilling the right to water. Environmental law and policy in general, and sustainable development specifically, demand the protection of the environment and thereby contribute to the fulfilment of the human right to water. These regulations govern the quality and quantity of the water which is available and the ways in which water is managed. Figure 5.1 illustrates this relationship between the human right to water and the indirect obligations. The human right to water can be considered as at the core.

This chapter analyses which aspects of sustainable development as explained in chapter 4 can strengthen the conceptualisation and further development of the human right to water. It also explains where possible tensions may arise which need to be reconciled. Integrating human rights and sustainable development provides additional legal reasoning for state obligations in relation to access to water for domestic

[731] Stephanie Chuffart and Jorge E Vinuales, 'From the Other Shore: Economic, Social and Cultural Rights from an International Environmental Law Perspective' in Eibe Reidel and Christophe Golay (eds), *Economic, Social and Cultural Rights in International Law* (Oxford University Press 2014). p. 6.

purposes.[732] In addition, sustainable development can elucidate those elements of the human right to water which need further development and adds guidance for implementation by states.

This chapter starts by discussing (5.2) the often mentioned paradox between sustainability and universal access. This is followed by an analysis of the compatibilities of the human right to water and sustainable development and their differences (5.3). Section 5.4 then discusses the prioritisation of water for domestic purposes within the water allocation process. It looks at the provisions within international water law and environmental law that mention which priorities states should set in relation to water management and analyses whether there is support for prioritising access for domestic and personal uses. The realisation of the human right to water for vulnerable groups in relation to the management of water resources may demand a reallocation of water for domestic purposes and the protection of water resources. The subsections of section 5.4 discuss the links between sustainable development and the obligation for states to progressively realise the right to water using the maximum available resources, the principle of non-retrogression and the duty of states to ensure international cooperation and assistance. Section 5.5 deals with the minimum core obligations and relates them to sustainable development. Sustainable development is not only a requirement for progressively realising the right but also for the minimum standard of the right. The subsections of section 5.6 discuss other aspects of sustainable development which are relevant for the human right to water, such as the duty to ensure the sustainable use of natural resources and integration. Section 5.7 delves into the procedural aspects where the coming together of the two fields provides additional remedies for the protection of access to water. Finally, section 5.8 summarises how the human right to water and sustainable development can be integrated.

5.2 THE EVOLVING RELATIONSHIP BETWEEN THE HUMAN RIGHT TO WATER AND SUSTAINABLE DEVELOPMENT

Having explored the principles of the 1992 Rio Declaration and the 2002 Johannesburg Summit in the previous chapter, it has become clear that comprehending why complementing human rights with sustainable development in general can be a beneficial approach especially for the human right to water. Human rights and sustainable development are complementary and they can be mutually reinforcing and incorporate social, political, economic and cultural aspects.[733] Some also argue that

[732] Vasilka Sancin and Masa Kovic Dine, 'Ensuring Access to Safe Drinking Water as an Imperative of Sustainable Development' in Volker Mauerhofer (ed), *Legal Aspects of Sustainable Development: Horizontal and Sectorial Policy Issues* (Springer International Publishing Switserland 2016). p. 95.

[733] Milena Petters Melo, 'Efficiency, Sustainability, and Justice to Future Generations' in Klaus Mathis (ed), *Efficiency, Sustainability, and Justice to Future Generations*, vol 98 (Springer Netherlands 2011) <http://link.springer.com/10.1007/978-94-007-1869-2> accessed 18 September 2013. p. 139.

the obligation of sustainable development includes the obligation for states to provide access to safe drinking water. This is because sustainable development comprises the obligation to ensure the availability of water resources.[734]

The inclusion of sustainability and sustainable development[735] in the human right to water is supported by General Comment No. 15 which mentions: 'the manner of the realisation of the right to water must also be sustainable, ensuring that the right can be realised for present and future generations.'[736] Even though the Committee mentions sustainable access several times in this Comment[737], it does not elaborate what sustainable access entails. However, it refers to the definitions of sustainability and sustainable development as discussed in the Rio Declaration.[738] In 2013, the UN Special Rapporteur on the human right to safe drinking water and sanitation offered additional insight into the relevance of sustainability for the human right to water. In this report she argues for the inclusion of sustainability as a principle of human rights.[739] This report primarily deals with economic considerations such as budgeting and economic growth and the connection between sustainability and core concepts of human rights.[740] However, there is still a need for clarification concerning sustainability in relation to this right.

These two documents, General Comment No. 15 and the report of the Special Rapporteur, were not adopted by states. Therefore it is difficult to argue that the inclusion of sustainability as part of the human right to water is supported by states and will become part of their implementation efforts. However, support is increasing. And there are international documents already in force which can support efforts for realising this right. With the adoption of the Sustainable Development Goals in 2015, states also accepted Goal No. 6 on ensuring access to water and sanitation for all, with one of the targets being ensuring sustainable withdrawals and the supply of freshwater.[741]

---

734 (Sancin) p. 99.

735 When it comes to the use of the terms sustainable development and sustainability. This study adopts the perspective that sustainability should be seen as the long-term end goal while sustainable development has shaped the processes which need to take place in order to attain sustainability. Therefore in order for a sustainable human right to water to be realized, sustainable development can be used as an instrument and guide the steps that will need to be taken in order to reach the end goal.

736 (Committee on Economic, Social and Cultural Rights, 2002) par. 11.

737 See (Committee on Economic, Social and Cultural Rights, 2002) paragraphs 26, 28,

738 See footnote 5 The Committee refers to the Rio Declaration and Agenda 21 for a definition of sustainability. Footnote 11: 'For a definition of sustainability, see the Report of the United Nations Conference on Environment and Development, Rio de Janeiro, 3–14 1992, Declaration on Environment and Development, principles 1, 8, 9, 10, 12 and 15; and Agenda 21, in particular principles 5.3, 7.27, 7.28, 7.35, 7.39, 7.41, 18.3, 18.8, 18.35, 18.40, 18.48, 18.50, 18.59 and 18.68.'
Footnote 22: Agenda 21, chaps. 5,7 and 18; and the World Summit on Sustainable Development, Plan of Implementation (2002), paras 6 (a), (l) and (m), 7, 36 and 38.
The Committee mentions sustainable access to water in paragraphs 11, 26 and 28.

739 De Albuquerque, 'Report of the Special Rapporteur on the Human Right to Safe Drinking Water and Sanitation'. paragraph 9.

740 Ibid.
The considerations of the Special Rapporteur will be discussed in the following chapter.

741 Sustainable Development Goals, goal no. 6.

Sustainable development and the human right to water are two concepts which can be converged. This idea is not new as the two concepts have had both strong and weak ties during the past few decades. Access to water and sustainable development have been closely connected in the international arena. Access to water started off by taking environmental concerns into account. However, after experiencing several developments it seems to have lost these concerns while they are highly relevant in the current context with increasing water scarcity. As explained previously, access to water has often been part of many international documents in relation to the environment and sustainable development. What can be observed is that access to water has moved from the field of environmental management and water law into the domain of human rights. Along the way its articulation has lost certain properties. The 1990 New Delhi Statement[742] mentions the elements of availability, quality and affordability similar to General Comment No. 15. It also states that one of the policy aims is the following:

> *'To implement programmes aimed at expanding service coverage within the framework of integrated water resources and environmental planning and management, in the context of sustainable national social and economic plans and urban and rural development policies, and to orient them towards services that reflect community needs and are used by beneficiaries'*[743]

After the New Delhi Statement additional non-binding international declaration documents followed which placed emphasis on access to water in relation to environmental concerns or sustainable development such as the Dublin Statement on Water and Sustainable Development[744] and the Bonn International Conference on Freshwater Ministerial Declaration.[745] However, the most recent declarations on access to water *within human rights* lack the consideration of environmental concerns. The right as recognised by the Human Rights Council only recognises the right to safe and clean drinking water, similar to the 2010 GA Resolution.[746] Nevertheless, the HRC has shifted towards a more environmental interpretation. In its 2014 Resolution the HRC refers to the Open Working Group on Sustainable Development Goals and highlights the element of sustainability.[747] It also calls upon states to achieve sustainable universal access.[748]

742 New Delhi Statement, Global Consultation on Safe Water and Sanitation www.un.org/documents/ga/res/45/a45r181.htm.

743 4(b)

744 Dublin Statement on Water and Sustainable Development, International Conference on Water and Environment, 31 January 1992.

745 Bonn Ministerial Declaration, Conference Report, International Conference on Freshwater, Bonn, 3–7 December 2001 and Bonn Recommendations for Action, Conference Report, International Conference on Freshwater, Bonn, 3–7 December 2001.

746 UN General Assembly Resolution A/RES/64/292.

747 Article 4 A/HRC/27/7 https://documents-dds-ny.un.org/doc/UNDOC/GEN/G14/177/87/PDF/G1417787.pdf?OpenElement.

748 Article 11b.

With the formulation of the Sustainable Development Goals this track was continued and access to water again takes environmental concerns into account. Target 6.1 explains that by 2030 universal and equitable access to safe and affordable drinking water for all should be achieved. The connected targets describe the steps which need to be taken in order to achieve the main target and the majority of these targets focus on the environment.[749] Target 6.3 for instance mentions: 'By 2030, improve water quality by reducing pollution, eliminating dumping and minimising release of hazardous chemicals and materials, halving the proportion of untreated wastewater and substantially increasing recycling and safe reuse globally.'[750]

From the above it can be concluded that in the international arena access to water began as part of international environmental law and water law and transitioned into a human right, while losing some particularities along the way, which included environmental concerns while these aspects are highly relevant for the effective realisation of the human right to water. Nevertheless, one of the benefits of access to water being recognised as a human right is the need for it to be a right with certain entitlements. It seems, however, that sustainability and the human right to water are once again converging and this may be the future trend in the development of this right. What needs to be explained is what sustainable development actually means for the human right to water. This will be further explained in this chapter.

5.3 OVERLAPS, DIFFERENCES AND TENSIONS

In order for water resources to be available for the realisation of vulnerable individuals' human right to water, states need to take into account several requirements regarding the priority of use, the allocation or reallocation of water resources, the protection of water resources against pollution and depletion, the protection of ecosystems and a

---

[749] 6.1 By 2030, achieve universal and equitable access to safe and affordable drinking water for all
6.2 By 2030, achieve access to adequate and equitable sanitation and hygiene for all and end open defecation, paying special attention to the needs of women and girls and those in vulnerable situations
6.3 By 2030, improve water quality by reducing pollution, eliminating dumping and minimizing release of hazardous chemicals and materials, halving the proportion of untreated wastewater and substantially increasing recycling and safe reuse globally
6.4 By 2030, substantially increase water-use efficiency across all sectors and ensure sustainable withdrawals and supply of freshwater to address water scarcity and substantially reduce the number of people suffering from water scarcity
6.5 By 2030, implement integrated water resources management at all levels, including through transboundary cooperation as appropriate
6.6 By 2020, protect and restore water-related ecosystems, including mountains, forests, wetlands, rivers, aquifers and lakes
6.a By 2030, expand international cooperation and capacity-building support to developing countries in water- and sanitation-related activities and programmes, including water harvesting, desalination, water efficiency, wastewater treatment, recycling and reuse technologies
6.b Support and strengthen the participation of local communities in improving water and sanitation management.

[750] Target 6.3.

number of procedural obligations. These requirements are included within sustainable development and assist in attaining sustainability. In this way, the two concepts complement each other.

A key difference between the concepts is the different perspectives they take on water. The most authoritative interpretation of the human right to water falls within the scope of the Covenant on Economic, Social and Cultural Rights, clarifying that the right has economic, social and cultural aspects. Sustainable development has however identified water as an ecological, economic and social asset.[751] As shown in Chapter 4, sustainable development has a tendency to prioritise its economic dimensions, focusing on the economic growth of the state.

Another difference and a possible tension between sustainable development and the human right to water is the difference in their scope and objective. The human right to water as currently developed adopts an individual approach while sustainable development adopts a collective approach and requires states to implement comprehensive solutions which would benefit not only individuals but communities and future generations both in those states and even beyond.[752] Sustainable development provisions are often aimed at states and do not create any entitlement for individuals. In this sense the human right to water will allow for sustainable development to become more justiciable.

The objectives of sustainable development and human rights are the same, as both aim to improve human conditions. Sustainable development is human-centred and is aimed at safeguarding a high quality of life for current and future generations. And in order for this quality of life to be realised the fundamental needs of people need to be fulfilled. Human rights take a similar approach as they aim to recognise human dignity and are meant to guarantee a minimum standard of living for individuals facilitated by the state.

Even though they share the same goals, they may emphasise different aspects. One of the considerations which both fields lack is the protection and preservation of the environment for its own intrinsic value. Both human rights as well as sustainable development consider the environment as a resource for humans to use.[753] Within human rights it is recognised that human rights and the protection of the environment are connected, given that the realisation of many rights depends on natural resources. Because of this, the idea underlying a human right to the environment started in the 1990s. However, the articulation of this right has stalled, as the UN Draft Principles on Human Rights and the Environment have not been formally adopted.[754] These Draft Principles include Article 6 which declares that: 'all persons have the right to protection

751 CoE Johannesburg World Summit on Sustainable Development, preamble.

752 Sancin and Dine. p. 97.

753 Maria Adebowale and others, 'Environment and Human Rights: A New Approach to Sustainable Development' [2001] Opinion World Summit on Sustainable Development 1 <http://pubs.iied.org/pdfs/11016IIED.pdf>. p.4.

754 Richard P. Hiskes. p. 100.

and preservation of the air, soil, water, sea-ice, flora and fauna, and the essential processes and areas necessary to maintain biological diversity and ecosystems.' This shows that the preservation of ecosystems is tied to humans and is not considered independently.[755]

For sustainable development Principle 1 of the Stockholm Declaration represents the same notion, as it states:

> *Man has the fundamental right to freedom, equality and adequate conditions of life, in an environment of a quality that permits a life of dignity and well-being, and he bears a solemn responsibility to protect and improve the environment for present and future generations.*

In some regional and national contexts, the environment is increasingly recognised for its intrinsic value and is even given certain rights. The Constitution of Ecuador grants nature a 'right to exist, persist, maintain and regenerate its vital cycles, structure, functions and its processes in evolution.'[756] Other prime examples can be found in relation to water.[757] In 2012, New Zealand recognised the rights of the Whanganui River.[758] And South Africa became the first country to formally enact environmental water requirements in its legislation by including the ecological reserve in its National Water Act.[759] 'The ecological reserve relates to the water required to protect the aquatic ecosystems of the water resource. The Reserve refers to both the quantity and quality of the water in the resource, and will vary depending on the class of the resource.'[760]

Human rights and sustainable development are centred on human beings which makes them compatible in that sense and as a result it can be argued that the environment will most likely be sacrificed whenever there is a conflict with human interests such as economic development.[761] However, due to the current water crisis it can be imagined that an eco-centric approach might become necessary under certain circumstances. For instance, reallocating water to certain areas may become ill-advised if this would put an irreversible strain on resources.

---

755 Bourquain. p. 70.

756 Art. 71.

757 See also Liping Dai, 'A New Perspective on Water Governance in China: Captain of the River' (2015) 40 Water International 87 <www.tandfonline.com/doi/abs/10.1080/02508060.2014.986702>

758 Louis J Kotze, 'Human Rights, the Environmental, and the Global South', *International Environmental Law and the Global South* (Cambridge University Press 2015). p. 180. See Jennifer L Archer, 'Rivers, Rights & Reconciliation in British Colombia: Lessons Learned from New Zealand's Whanganui River Agreement' (2014).

759 Harrison Pienaar, Antonia Belcher and Dana F Grobler, 'Protecting Aquatic Ecosystem Health for Sustainable Use' in Barbara Schreiner and Rashid Hassan (eds), *Transforming Water Management in South Africa: Designing and Implementing a New Policy Framework* (Springer Science 2011) <http://link.springer.com/10.1007/978-90-481-9367-7>. p. 126.

760 Part 3 https://www.dwa.gov.za/Documents/Legislature/nw_act/NWA.pdf.

761 Bourquain. p. 70.

5.4 THE OUTER BOUNDARIES OF THE HUMAN RIGHT TO WATER

As explained there are obligations stemming from sustainable development which support the human right to water and which define the outer boundaries of this human right to water. This right, as mentioned before, has to function in a particular physical context. This section discusses the furthest outer boundary of the human right to water, namely the allocation of water resources between states which is regulated, amongst others, by the principles of international cooperation and common but differentiated responsibilities. These principles are derived from different instruments.

5.4.1 INTERNATIONAL COOPERATION AND COMMON BUT DIFFERENTIATED RESPONSIBILITIES

So far, it has been explained that primary responsibility for realising the human right to water for all individuals lies with the state and in order to fulfil this responsibility it has to use all available resources, including water resources. And it has also been explained that states are not allowed to take retrogressive measures. Even so, there may be situations where states do not have enough water resources available to meet their obligations. According to the Human Development Report, climate change will negatively affect water run-off in many regions.[762] And by 2050 it is estimated that more than half of the world's population will live in water-stressed areas.[763] These projections show that states will indeed be confronted with water scarcity which will influence the availability of water.

Under these circumstances, there are obligations for other states to provide assistance to states which are experiencing water stress and for this reason are unable to provide individuals with the amount of water that the human right to water guarantees. 'Depending on the availability of resources, States should facilitate realisation of the right to water in other countries, for example through provision of water resources, financial and technical assistance, and provide the necessary aid when required.'[764] The type of assistance depends on geographical proximity and whether the helping and receiving states share water resources.[765]

Under this obligation, a distinction can be made between two different but connected requirements in which sustainable development can inform the human right to water.

---

762 See United Nations Development Programme. p. 162 on the change in water run-off compared with the 1961–1990 average (%) by 2050 based on IPCC scenario A.

763 CA Schlosser and others, 'The Future of Global Water Stress: An Integrated Assessment', vol 2 (2014) <http://onlinelibrary.wiley.com/doi/10.1002/2014EF000238/abstract>. p. 27.

764 (Committee on Economic, Social and Cultural Rights, 2002) paragraph 34.

765 Bulto, *The Extraterritorial Application of the Human Right to Water in Africa*. p. 124. In addition, providing assistance is also necessary for meeting cultural requirements. (Committee on Economic, Social and Cultural Rights, 2002) paragraph 34.

Firstly, the obligation of states to provide assistance to states which do not have enough water resources for realising the right to water for their citizens. Before international assistance can be provided by individual states or in cooperation, receiving states must demonstrate that they have used nationally available freshwater resources to the maximum extent and have prioritised its allocation for domestic and personal uses.[766] For instance, the Republic of Yemen is currently experiencing increasing water scarcity. One of the reasons for this is the allocation of water resources for the production of khat (a plant whose leaves are used as a stimulant), which is a water-intensive crop.[767] This allocation of scarce resources could challenge states' willingness to provide assistance when Yemen is expected to run out of water by 2050.

Secondly, there is a prohibition on states taking action which prevents other states from realising the right to water for their citizens by, for instance, overexploiting shared water resources, thereby restricting other states in their access to water and this can be caused by the activities of co-riparian states such as pollution, diversion or overconsumption or building dams.[768] The human right to water also requires states to refrain from actions which directly or indirectly limit other states in fulfilling this right for their inhabitants.[769] And this is supported by international water law which requires states to equitably share water resources for the benefit of their populations.[770] International law and international water law can therefore specifically complement this requirement of the human right to water and thereby strengthen its position in international law as well as its further development. International water law assists by ensuring that states have the amount of water that they need while the human right to water determines the division of water within the boundaries of the state by prioritising water for domestic and personal uses.

It has now been established that there is a duty under both human rights law as well as international water law for co-riparian states to share freshwater resources in such a way that those states are able to foresee the basic needs of their inhabitants. What should be kept in mind, however, is that General Comment No. 15, the Draft Articles on the law of the non-navigational uses of international watercourses and the Watercourses Convention only refer to this duty as an interstate affair. This means that individuals do not have a right to any redress when the duty to assist has been violated by assisting states. In addition, international human rights law is still lacking when it comes to extraterritorial obligations which allow individuals in one state to bring a claim against the government of another state.[771] Extraterritorial obligations would provide redress in those situations where states externalise pollution or engage in unsustainable usage at the expense of individuals in other states.[772]

---

766 (Committee on Economic, Social and Cultural Rights, 2002) paragraph 41.

767 Misiedjan 2015, p. 203.

768 Bulto, *The Extraterritorial Application of the Human Right to Water in Africa*. p. 122.

769 (Committee on Economic, Social and Cultural Rights, 2002) paragraph 31.

770 Vink, Takeuchi and Kibler. p. 10.

771 McCaffrey. p. 20 Maastricht Guidelines.

772 Ibid. p. 20.

Another challenge posed by the duty to provide assistance when sharing watercourses is balancing the pillars of sustainable development. The constraint that water for basic needs is to be prioritised over other water uses also affects how the duty of international cooperation is fulfilled. States which, for instance, have plans to develop their water resources in a way which will negatively affect the quality and quantity of freshwater resources in other states and thereby threaten the human right to water of individuals in those other states are also required to consider this prioritisation.[773] Economic development in one state should not take priority over the realisation of the human right to water in another state.

In order to assist states in meeting their duty of international cooperation, transparency, accountability and communication between riparian states will be necessary. Receiving states will have to notify assisting states when national freshwater resources are at risk of falling below what is necessary for providing their inhabitants with necessary access.[774]

Lastly, one of the challenges in fulfilling this duty is identifying how much water should be made available for receiving states. What should first be explained is that this depends on the freshwater resources that the assisting state has available, especially when this state is still required to meet its duty towards its own inhabitants. Nonetheless, when resources are available, it can be argued that the receiving state is entitled to receive enough water to fulfil the minimum core requirement. The principle of equitable and reasonable utilisation could be applied as a formula for distribution which will mean that the receiving state is not entitled to receive more water from the co-riparian state than the equitable and reasonable use of the international watercourse demands.[775] One problem is that the principle of equitable utilisation applies exclusively to international watercourses, meaning transboundary water courses. Therefore not all water resources are covered by this principle and shared groundwater resources are not covered by a legally binding international instrument.[776] However, several general principles of sustainable development apply to groundwater management.[777] For instance the principle of rational utilisation can be applied.[778] Rational utilisation is directly connected to these elements as the availability and quality of water resources determine the scope for water use.[779]

The above shows that there is an obligation for states to provide necessary assistance to other states which do not have the necessary resources for the realisation of the

773 (McCaffrey, 1992) p. 22.

774 Bulto, *The Extraterritorial Application of the Human Right to Water in Africa*. p. 124.

775 McCaffrey, 1992) p. 23.

776 See Joseph W Dellapenna, 'The Customary Law Applicable to Internationally Shared Groundwater' (2011) 36 Water International 584 <www.tandfonline.com/doi/abs/10.1080/02508060.2011.599025>; Kirstin I Conti and Joyeeta Gupta, 'Global Governance Principles for the Sustainable Development of Groundwater Resources' (2016) 16 International Environmental Agreements: Politics, Law and Economics 849.

777 Ibid. p. 865.

778 Dellapenna, 'The Customary Law Applicable to Internationally Shared Groundwater'. p. 586.

779 Sancin and Dine. p. 101.

human right to water. This assistance can be provided by adjusting the use of shared watercourses but also by the sharing of technology and providing financial assistance. The inclusion of the perspective of international water law provides a direct obligation for co-riparian states that the human rights framework lacks and the human right to water regulates the division of water resources within the state.

The principle of common but differentiated responsibilities can easily be linked to the duty to receive international assistance and cooperation and the duty to progressively realise the human right to water. The principle of common but differentiated responsibilities contains two premises. It considers the contributions of states to environmental problems and it also considers the capacity of states to address these problems. These three elements take into account the capacity of states to meet their obligations.

One of the differences is however that the principle of common but differentiated responsibilities is motivated by the notions of social justice and distributive justice. Those which have contributed most to problems such as climate change[780] are expected to make the greatest contributions for solving these problem.[781] The principle also emphasises solidarity. This could mean that in relation to water consumption, those who have contributed most to water becoming scarcer should assist those countries which are experiencing water scarcity as a result of their actions.[782]

Interesting in this context is that many developed states which abstained from voting on the 2010 GA Resolution on the human right to water and sanitation were concerned about its effect. Recognising water as a human right to water may have consequences concerning their obligations to provide assistance to developing states in order to realise this right for their citizens. For instance, the Netherlands stated that 'the text placed insufficient responsibility on national Governments, upon which citizens must be able to rely and from which they must obtain redress.'[783] General Comment No. 15 nevertheless placed special responsibility on developed states in assisting poorer developing states.[784]

This is in line with the obligation in human rights law where states are expected to provide each other with assistance and cooperation. Within sustainable development this idea is also mentioned. The Implementation Plan of the World Summit on Sustainable Development refers to Principle 7 of the Rio Declaration which mentions the

---

780 B Muller, 'Varieties of Distributive Justice in Climate Change' (2001) 48 Climate Change 273 <http://jmq.sagepub.com/lookup/doi/10.1177/107769900908600101>.

781 Yoshiro Matsui, 'Some Aspects of the Principle of " Common but Differentiated Responsibilities "' (2002) 2 International Environmental Agreements: Politics, Law and Economics p. 152.

782 Sara De Vido, 'The Right to Water as an International Custom : The Implications in Climate Change Adaptation Measures' (2012) 3 Carbon and Climate Law Review p. 226.

783 Www.un.org/News/Press/docs/2010/ga10967.doc.htm. Also see: H. F. M. W. Van Rijswick, 'Improving the Right to Water in the Netherlands' in Henri Smets (ed), *The implementation of the right to safe drinking water and sanitation in Europe* (2011).

784 Committee on Economic Social and Cultural Rights. Paragraph 34.

principle of common but differentiated responsibilities[785] and relates it to interrupting and changing unsustainable patterns of consumption and production. Even though states bear the main responsibility for realising sustainable development within their own countries, other states are encouraged to take measures which support developing countries in doing this.[786] Developing states are therefore placed at the forefront.

Within the human rights framework it is often challenging to determine whether states have met their obligation to progressively realise rights and use all available resources accordingly. The focus in the human rights framework lies on the examination of fiscal and policy efforts.[787] States are required to demonstrate that they have made maximum efforts to realise social rights, which has been explained in a previous section. Besides strengthening accountability, it can also be imagined that when states can demonstrate, under these circumstances, that national resources are lacking, other states may feel morally bound to meet the duty to provide assistance.[788]

5.4.2 PRIORITY OF USE

It is often argued that providing access to those who currently do not have access to water will increase environmental degradation and the depletion of water resources.[789] However, as mentioned in Chapter 1, on a global basis household consumption accounts for less than 10 percent of the overall human water use as is illustrated in Figure 5.2.[790] Because of this, household consumption does not pose a major threat to water resources.[791] Also, there is sufficient water to meet the basic needs of everyone. Even in countries where water is limited there is still enough to meet basic needs.[792] The issue of who gets water and how much is not exclusively an issue of the availability of

---

785 Report of the United Nations Conference on Environment and Development, Rio de Janeiro, 3–14 June 1992 (A/CONF.151/26/Rev.1.

786 James Heenan, Hossein Fadaei and Simon Walker, 'Sustainable Development' [2002] NGLS Roundup 1. p.8.

787 See Sakiko Fukuda-parr and Terra Lawson-remer, 'Measuring the Progressive Realization of Human Rights Obligations : An Index of Economic and Social Rights Fulfillment Department of Economics Working Paper Series' (2008) Paper 2008 Economics Working Papers 1.

788 Michael R Ulrich, 'The Impact of Law on the Right to Water and Adding Normative Change to the Global Agenda' 48 George Washington International Law Review p. 51.

789 See for instance Trigueros, (2012) where it is argued that: To answer the questions posed in the introduction: the considerably weakened state of international environmental law, following the shift in focus from environmental protection to a right to development, coupled with the potential realization of a human right to water could very well lead to rapid depletion of the planet's freshwater resources (p. 625)

790 "Water for the People, Water for Life" United Nations World Water Development Report UNESCO 2003.

791 United Nations Development Programme. p. 3.

792 Winkler, *The Human Right to Water Significance, Legal Status and Implications for Water Allocation.* p. 7; (Khalfan et al., 2008) p. 87.
According to Khalfan and others.: Only in a few regions of a few countries would there be insufficient water resources for essential domestic purposes such as drinking, washing (clothes and personal hygiene), cooking and sanitation – if such domestic uses were prioritised over other uses.

water resources. As the Human Development Report explains: 'the scarcity at the heart of the global water crisis is rooted in power, poverty and inequality, not in physical availability.'[793] Ensuring access to water for vulnerable groups is therefore also a question of prioritisation, reallocation, combating inequality and promoting empowerment.

Figure 5.2 Competing water uses for main income groups of countries

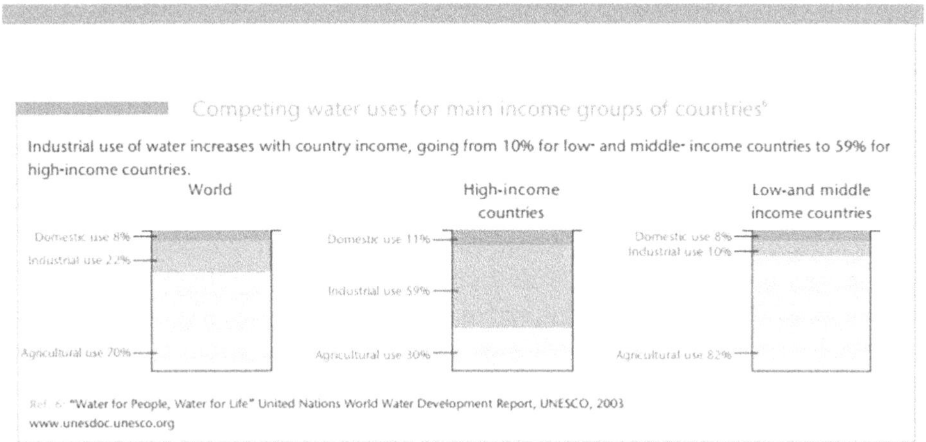

There are sufficient water resources to meet the basic needs of all individuals, yet a large section of the world population does not have access to water for fulfilling their needs. If a more equitable distribution of water resources is to be achieved and the basic needs of all, including vulnerable groups, are to be met, water resources should be reallocated and priority should be given to domestic and personal uses.[794] This section discusses whether the human right to water allows the allocation of water for domestic purposes to be prioritised. It also discusses provisions on prioritisation in environmental law and water law and analyses whether they can provide an additional line of reasoning for prioritising the allocation of water for basic needs. Finally, this section advocates a basic prioritisation structure.

Priority of use entails that in the event of a conflict between water uses or other interests, one use or interest will first receive the amount of water that the priority establishes.[795] For example, in relation to the human right to water it could be said that of all domestic uses that are covered by this right, access to water in order to prevent dehydration and diseases has priority[796] over other domestic uses such as: the washing of clothing, food preparation and household hygiene.[797] However, in relation to other

[793] United Nations Development Programme. p.2.
[794] Ibid. p. 3.
[795] Hildering. p. 100.
[796] (Committee on Economic, Social and Cultural Rights, 2002) paragraph 6 Winkler, *The Human Right to Water*.
[797] (Committee on Economic, Social and Cultural Rights, 2002) paragraph 12.

water uses (such as water for agriculture), identifying access to water for domestic and personal uses as a human right supports the claim of giving priority to allocating water for its realisation.[798] This classification allows for other water uses to be overridden when different uses are weighed against each other in policy making or through other avenues. This is important considering that currently only 10% of water globally is used for domestic purposes. Some of the remaining 90% of water use must be reallocated for domestic uses, especially to meet the needs of those who are currently unserved and vulnerable.

As the allocation of water is not only regulated by human rights but also by international water policy and law and environmental law, having these fields prioritise water for domestic and personal uses would support the realisation of the human right to water. If this is not the case, states might find it challenging to meet their obligations under these different systems. The first reference to prioritising water for domestic purposes can be found in the document which is considered to be the 'mother document' of the right to water: the 1977 Mara del Plata Action Plan. It mentions that in national development policies and plans priority should be given to supplying the entire population with drinking water, with the aim of a just and equitable distribution among the people.[799]

However, the Action Plan was followed by the Dublin Statement which highlighted the economic aspect of water. Principle No. 4 of the Dublin Statement reads: 'Water has an economic value in all its competing uses and should be recognised as an economic good.' This principle is relevant in prioritisation as it is seen as a tool for the wasteful and damaging use of water.[800] Focusing on economic value and efficiency in relation to water use runs the risk of overlooking the rights of vulnerable groups and individuals.[801] The human right to water obliges states to allocate water for domestic purposes, thereby recognising the cultural and social aspects of water and not having its economic aspect placed at the forefront.[802]

Another important instrument is the 1997 Watercourses Convention which includes Article 10:

1. *In the absence of agreement or custom to the contrary, no use of an international watercourse enjoys inherent priority over other uses.*
2. *In the event of a conflict between uses of an international watercourse, it shall be resolved with reference to Articles 5 to 7, with special regard being given to the requirements of vital human needs.*

---

798 (Committee on Economic, Social and Cultural Rights, 2002) paragraph 6: Nevertheless, priority in the allocation of water must be given to the right to water for personal and domestic uses. Priority should also be given to the water resources required to prevent starvation and disease, as well as water required to meet the core obligations of each of the Covenant rights.

799 P. 67.

800 See Dublin Statement p. 2.

801 Winkler, *The Human Right to Water Significance, Legal Status and Implications for Water Allocation.* p. 144.

802 (Committee on Economic, Social and Cultural Rights, 2002) paragraph 11.

This article states that no water use enjoys inherent priority over other uses. The article does however allow for special attention to be given to water for vital human needs in the event of a conflict between water uses. Nevertheless, vital human needs are not formulated as a priority. Having said that, the Watercourses Convention does leave room for the recognition of priority by agreement or custom. If the human right to water becomes part of custom, this Convention will grant it priority. Some argue that access to water deriving from the right to life under the ICCPR is already a part of customary international law.[803]

The Statement of Understanding of the Watercourses Convention also clarifies that '*in determining "vital human needs", special attention is to be paid to providing sufficient water to sustain human life, including both drinking water and water required for production of food in order to prevent starvation*'.

Another argument in favour of the prioritization of human needs under the Watercourses Convention can be found in Article 5:

> '*Watercourse States shall in their respective territories utilize an international watercourse in an equitable and reasonable manner. In particular, an international watercourse shall be used and developed by watercourse States with a view to attaining optimal and sustainable utilization thereof and benefits therefrom, taking into account the interests of the watercourse States concerned, consistent with adequate protection of the watercourse.*'

It is explained in the ILC Commentary to the Draft Convention that those uses which cause significant harm to human health and safety are intrinsically inequitable and unreasonable.[804]

According to the Brundtland Commission priority should be given to 'the essential needs of the world's poor.'[805] There are two conclusions which can be derived from the Conventions relating to the priority of use. First, access to water for basic needs is referred to as a need and not a right. Secondly, neither the Watercourses Convention nor the UNECE Convention grants priority to vital human needs.

In addition to these provisions there are some other instruments which support prioritisation for basic needs. Chapter 18 of the 1992 Agenda 21 underscores that priority should be given to allocating water for basic needs and for safeguarding ecosystems.[806] It argues that the multi-sectoral nature of water should be taken into account by using an Integrated Water Resources Management approach.[807] Paragraph 18 of the 2002 Johannesburg Declaration requires states to 'speedily increase access to such basic requirements as clean water, sanitation.'[808] The 2002 Implementation Plan of

803 See chapter 3 section …

804 ILC (1996), Commentary to Article 7, para. 14. In line with this Commentary, Nelissen (2002), p. 19, in relation to the use of internationally shared water resources, argues in favour of providing absolute priority to drinking water. McCaffrey.

805 Brundtland Commission. p. 41 under IV. Conclusion.

806 Agenda 21 Chapter 18.

807 Agenda??? paragraph 18.8.

808 Johannesburg Declaration on Sustainable Development A/CONF.199/20.

the World Summit on Sustainable Development also 'gives priority to the satisfaction of basic human needs and balances the requirement of preserving or restoring ecosystems.'[809]

These influential declarations on international water and environmental policy demonstrate that the prioritisation of basic human needs such as water for drinking is often considered. However, the question remains whether a general prioritisation of water for basic human needs is sufficient. This is because in the event of water scarcity, less water will be available for other water uses such as for agriculture which can compromise the realisation of the right to food.[810] The provisions on the prioritisation of basic needs need to be further explained. What needs to become clear is what falls under the notion of a basic need.[811] For instance, does it include water for sanitation or merely drinking water? And how much in quantity should be assigned per community or individual? Examples can be found in national legislation containing prioritisation provisions. States such as South Africa and Burkina Faso go beyond an unconditional general provision.[812]

The first priority, according to the majority of the above-mentioned instruments, should be given to domestic and personal uses. Following this, priority should be given to the allocation of water for subsistence farming and basic agriculture. General Comment No. 15 explains that:

> *The Committee notes the importance of ensuring sustainable access to water resources for agriculture to realize the right to adequate food (see General Comment No. 12 (1999)). Attention should be given to ensuring that disadvantaged and marginalized farmers, including women farmers, have equitable access to water and water management systems, including sustainable rain harvesting and irrigation technology. Taking note of the duty in Article 1, paragraph 2, of the Covenant, which provides that a people may not "be deprived of its means of subsistence", States parties should ensure that there is adequate access to water for subsistence farming and for securing the livelihoods of indigenous peoples.*[813]

However, beyond basic needs, prioritisation may become challenging especially when integrating sustainable development with the human right to water. This is because sustainable development entails the balancing of economic development, social development and environmental protection. The human right to water does not elaborate how states should shape their water management. The human right to water does entail that water should first be allocated for human survival, thereby guaranteeing basic health and avoiding thirst. This requirement is supported by

809 Paragraph 26 c.
810 Winkler, *The Human Right to Water Significance, Legal Status and Implications for Water Allocation.* p. 146.
811 Ibid. p. 147.
812 Ibid. p. 148 The South African Reserve establishes the Basic Human Needs Reserve, which sets aside a quantity of every single water body to be used for domestic purposes. It prioritizes only the essential needs of individuals which is consistent with the human right to water.
813 (Committee on Economic, Social and Cultural Rights, 2002) paragraph 7.

the right to life. However, at the national level states may also argue that in order to guarantee the survival of their inhabitants, allocating water for food production is required.[814] But also water uses such as energy and sanitation may be relevant at the survival level.[815] The national context of states determines which water uses are relevant for survival besides the basic needs of individuals. What is also relevant to consider is the availability of alternatives to water. Water resources should only be allocated for those uses for which water is essential. For instance, whether culturally acceptable and appropriate sanitation can be provided without using water.[816]

5.4.3 REALLOCATION AND ALLOCATION MEASURES

As has been established above, allocating water for domestic purposes has priority and will require a reallocation of water resources. Therefore different and new mechanisms should be implemented.[817] The reallocation of water resources will affect different water users. If, for instance, less water resources are distributed to agriculture, that sector will have to adjust its water use. This adjustment might require additional investment.[818] States should therefore consider compensation where necessary. What should also be considered are other measures which are not directly connected to the distribution of water but influence water use nonetheless. For example, in Yemen the subsidies for diesel encourage the use of pumping systems which extract large qualities of water.[819] Regulating these diesel subsidies differently could decrease the amount of water use.

Several states have implemented new allocation measures. South Africa is often seen as an example. As a general approach the South African government is aiming to redress historical inequalities between different ethnic groups stemming from the Apartheid system. This system of inequality also bled into water allocation. In the South African common law context, water rights are connected to land rights. Therefore in order to make more water available for domestic purposes land rights also need to be rearranged. South Africa has introduced a system of Compulsory Licensing (CL) to rearrange the existing legal access to water.[820] It is the main legal instrument mentioned in the National Water Act (NWA). The licensing system requires unlicensed users to reapply and those licensed are subject to review. 'The process involves a set of water resource management decisions, leading to a reallocation, with the primary intention

814 See the right to food in chapter 3.

815 Winkler, *The Human Right to Water Significance, Legal Status and Implications for Water Allocation.* p. 204.

816 The World Bank and others, *Sanitation for Development* (Second edi, 2008).

817 Van Rijswick and others. p. 8; Liping Dai, 'Identifying and Understanding the Main Challenges for Sustainable Water Resource Management in China' (2014) 24 Journal of Water Law 249.;

818 Van Rijswick and others. p. 8.

819 Misiedjan, van Rijswick and Tjen A Kwoei. p.199.

820 AJ Anderson and others, 'Integrated Monitoring of Water Allocation Reform in South Africa' (2008) 34 Water SA 731. p. 731.

of achieving race and gender equity, as well as to meet international obligations and the requirements of the Reserve.'[821]

The reallocation of water resources entails a complex puzzle which often starts with dividing water resources between states. When the water resources have been divided equitably between states, individual governments still have the responsibility to distribute these water resources fairly within their borders and to prioritise water for domestic and personal uses. This reorganisation requires an integral consideration of water uses and the stakeholders involved amongst other considerations.[822]

5.5 THE INNER CIRCLE: USING SUSTAINABLE DEVELOPMENT TO INTERPRET THE CORE OF THE HUMAN RIGHT TO WATER

The previous sections have discussed the outer boundaries of soft and hard provisions which support the implementation of the human right to water for vulnerable groups and individuals. The following sections delve into the elements of sustainable development which assist in articulating the human right to water. This section firstly discusses the substantive elements of the human right to water followed by the procedural elements and the rights to information, participation and a remedy.

5.5.1 SUBSTANTIVE ELEMENTS

The human right to water includes normative elements as explained in chapter 3. These elements, such as quality requirements, have already been further explored by organisations such as the World Health Organization (WHO). Nevertheless, there are other substantive elements which can use elements from sustainable development to further articulate the human right to water. The following sections explore general state obligations which apply to the human right to water. The incorporation of elements of sustainable development does not only support the further development of this specific right but also other rights which are connected to the environment.

5.5.1.1 Progressive realisation of the human right to water and core obligations

Progressive realisation is a central aspect relating to economic, social and cultural rights under international human rights treaties.[823] It entails that states should take steps towards reaching those obligations outside of the core and they have to be realised

821 Ibid. p. 731.
822 This will be further explored in the next chapter on the ten building blocks.
823 Progressive realization appears in international human rights treaties. Article 4 of the Convention on the Rights of the Child, Article 4 (2) of the Convention on the Rights of Persons with Disabilities and

immediately. The obligations of states therefore differ according to the immediacy of the realisation. The obligation to progressively realise applies beyond the core obligations which require immediate realisation. This section discusses the general obligation to realise the right to water progressively which is articulated in Article 2 paragraph 1 of the ICESCR:

> *Each State Party to the present Covenant undertakes to take steps, individually and through international assistance and co-operation, especially economic and technical, to the maximum of its available resources, with a view to achieving progressively the full realization of the rights recognized in the present Covenant by all appropriate means, including particularly the adoption of legislative measures.*

This provision limits the discretion of states when it comes to the implementation of economic, social and cultural rights. The following sections analyse these obligations and use the concept of sustainable development to inform interpretation.

5.5.1.2 The obligation to take steps to continuously improve conditions

Article 2(1) ICESCR is a specific characteristic of economic, social and cultural rights. It acknowledges that a full and immediate realisation of these rights is deemed to be difficult in a short period of time.[824] As General Comment No. 3 explains: 'It is on the one hand a necessary flexibility device, reflecting the realities of the real world and the difficulties involved for any country in ensuring full realisation of economic, social and cultural rights. On the other hand, the phrase must be read in the light of the overall objective, indeed the raison d'être, of the Covenant which is to establish clear obligations for States parties in respect of the full realisation of the rights in question.'[825] Nevertheless, there are clear obligations for states to meet. Regardless of their level of development, states must immediately take steps which are deliberate, concrete and targeted at the realisation of these rights.[826] Also states should progressively transition into implementing more sustainable ways of realising the right to water, allocating scarce resources more effectively and implementing more sustainable technologies.

5.5.1.3 Maximum available resources

The obligation to use the maximum available resources with the goal of progressive realisation is particularly interesting in relation to the human right to water. This obligation has mostly been discussed from an economic perspective, meaning the maximum available financial resources. States are expected to take into account

Article 1 of the Protocol of San Salvador. Therefore this discussion is also relevant for other economic, social and cultural rights instruments.

824 (Committee on Economic, Social and Cultural Rights, 1990) paragraph 9.

825 (Committee on Economic, Social and Cultural Rights, 1990) paragraph 9.

826 (Committee on Economic, Social and Cultural Rights, 1990) paragraph 2.

the financial resources which are necessary for the realisation of human rights. They are obliged to consider how to collect the needed finances and how to allocate these resources. However, in the case of the human right to water not only financial resources are relevant but also the natural resources on which the right is dependent: water. What needs to be established is which requirements the human right to water sets for the use of water resources with the aim of realising this right.

The connection between the human right to water and water as a natural resource seems to be straightforward. In this connection there is an opportunity to integrate human rights with environmental concerns. With a global phenomenon such as water scarcity due to climate change and the depletion of groundwater in mind, providing access to water may already become challenging in the near future and states might regress in their ability to realise the human right to water for individuals. For this reason it is crucial to consider how the sustainable use of natural resources and the realisation of the human right to water can be accomplished. Therefore this section will not only clarify the requirements for resource allocation which the human right to water sets but will also interpret the obligation to use the maximum available resources in line with the principle of the sustainable use of natural resources.

The level of state obligations differs per state and depends on the availability of resources. 'Resources' signify all means which are necessary and available for the realisation of these rights, not merely financial resources. Other important categories of resources which can be identified are the following: natural, human, organisational, technological, information and other resources.[827] It can therefore be said that states are required to determine which resources are most appropriate for meeting their obligation. Robertson illustrates this with the following:

> '*To the rural dweller, access to natural resources is of equal or greater significance than access to money. Giving a farmer land, water, seeds, and animals so he can fulfil his own right to food, is preferable to purely financial assistance, as it is more likely to alleviate the problem of perpetual state dependency.*'[828]

Sustainability should play an important role in this part of this core obligation. Using the maximum available resources can encourage an unsustainable use of water resources. Here a link can be made with the change that sustainable development has experienced, thereby increasing the importance of scientific knowledge to inform the allocation of scarce resources.[829] This aspect illustrates the human-centred approach that the human right to water has taken.

The use of water resources should not go beyond replenishment rates, if water resources are to be protected from depletion and exploited sustainably.[830] Surface water

---

827 Robertson. p. 697.
828 Ibid. p. 695.
829 Center for Strategic and International Studies, 'Global Water Futures. Addressing Our Global Water Future' (2005). p. 72.
830 Antonelli and Greco. p. 7.

and groundwater are both used to foresee water needs depending on its availability. Groundwater is currently a significant source of water for human consumption.[831] This source of water is convenient as relatively large volumes of water are stored and can be made available for withdrawal even when rainfall has been limited or absent. The risk attached to these sources is that they can be exploited to exhaustion if not controlled properly, making it scarce for future generations.[832] Global groundwater abstraction has increased drastically during the past 50 years. It has at least tripled and has increased by 1 to 2 percent annually.[833] Considering sustainable development as protection from depletion also means that states should consider which water resources should be used for which water uses. For instance, groundwater is often less contaminated than surface water and therefore does not require much filtration. However, extraction costs might be higher than for surface water. States should therefore consider cost-recovery measures.[834]

Water is also necessary to sustain different ecosystems. 'While efforts are being made to identify basic eco-system water requirements, there is little agreement about minimum water needs for the environment and few legal guarantees for environmental water have been set.'[835] States are forced to consider trade-offs regarding the maintenance and restoration of ecosystems.[836]

5.5.1.4 The obligation to abstain from taking deliberately retrogressive measures except under specific circumstances

States are required to move forward in the realisation of rights and not to take direct or indirect retrograde steps unless this is absolutely necessary. This is also referred to as the prohibition on retrogressive measures. The Committee explains that: 'any deliberately retrogressive measures in that regard would require the most careful consideration and would need to be fully justified by reference to the totality of the rights provided for in the Covenant and in the context of the full use of the maximum available resources.'[837] From this it can be concluded that the burden lies on states parties to prove that they have carefully considered all alternatives before taking any retrogressive steps, considering the totality of the rights in the Covenant and the maximum available resources.[838]

In relation to the realisation of the human right to water several comments can be made. Firstly, with water scarcity increasing, less water will be available which can

---

831 UNESCO. p. 85.
832 Antonelli and Greco. p. 7.
833 UNESCO. p. 84.
834 Hildering. p. 73.
835 Gleick, 'Water In Crisis : Paths To Sustainable Water Use'. p. 575.
836 (Gleick, 1998) p. 575.
837 (Committee on Economic, Social and Cultural Rights, 1990) paragraph 9.
838 Maastricht Guidelines on Violations of Economic, Social and Cultural Rights, Maastricht, Guideline No. 14.

lead to retrogression in the fulfilment of this right. However, this retrogression can be justified when the state can prove that it indeed does not have sufficient water resources available. In assessing this justification states should show that they have given careful consideration to any alternatives. This is relevant especially considering competing water uses. States will have to argue that reallocating water resources from other uses so that they can be directed towards domestic uses cannot be accomplished. They must show that even in the event of retrogressive measures, priority for the basic needs of the most vulnerable and marginalised groups is still given in national planning.[839]

5.5.1.5 *Relating the minimum core obligations of the human right to water to sustainability*

Outside the obligation of progressive realisation, the human right to water includes core obligations which need to be fulfilled immediately, also referred to as the minimum core.[840] The minimum core should be seen as the baseline from which progressive realisation starts.[841] General Comment No. 3 explains that: 'a minimum core obligation to ensure the satisfaction of, at the very least, minimum essential levels of each of the rights is incumbent upon every State party.' This means that core obligations have immediate effect and are meant to guarantee the minimum essential levels of the right.[842] General Comment No. 15 explains that the human right to water entails 9 core obligations:

> *(a) To ensure access to the minimum essential amount of water that is sufficient and safe for personal and domestic uses to prevent disease;*
> *(b) To ensure the right of access to water and water facilities and services on a non-discriminatory basis, especially for disadvantaged or marginalized groups;*
> *(c) To ensure physical access to water facilities or services that provide sufficient, safe and regular water; that have a sufficient number of water outlets to avoid prohibitive waiting times; and that are at a reasonable distance from the household;*
> *(d) To ensure personal security is not threatened when having to physically access to water;*
> *(e) To ensure equitable distribution of all available water facilities and services;*
> *(f) To adopt and implement a national water strategy and plan of action addressing the whole population; the strategy and plan of action should be devised, and periodically reviewed, on the basis of a participatory and transparent process; it should include methods, such as right to water indicators and benchmarks, by which progress can be closely monitored; the process by which the strategy and plan of action are devised, as well as their content, shall give particular attention to all disadvantaged or marginalized groups;*
> *(g) To monitor the extent of the realization, or the non-realization, of the right to water;*

---

839 Catarina De Albuquerque, 'Climate Change and the Human Rights to Water' (2010) <www.ohchr.org/Documents/Issues/Water/Climate_Change_Right_Water_Sanitation.pdf>. p. 10.

840 (Committee on Economic, Social and Cultural Rights, 2002) p. 37.

841 Chapman (2002) Core obligations: building a new framework for Economic, Social and Cultural Rights p. 9.

842 Committee on Economic and Social and Cultural Rights. paragraph 10.

(h) *To adopt relatively low-cost targeted water programmes to protect vulnerable and marginalized groups;*
(i) *To take measures to prevent, treat and control diseases linked to water, in particular ensuring access to adequate sanitation.*[843]

One of the questions could be whether sustainability is part of the core obligations or whether it only applies outside of the minimum core in relation to progressive realisation. What can be noticed, however, is that sustainability plays a role in each of these core obligations. It directly connects access to safe drinking water to sustainable development.[844] The environment needs to be protected in order to have the necessary quality and quantity of water available for the realisation of the right.

Access for individuals should be protected from disconnections and contamination. The Committee mentions in paragraph 10 that:

> '*The right to water contains both freedoms and entitlements. The freedoms include the right to maintain access to existing water supplies necessary for the right to water, and the right to be free from interference, such as the right to be free from arbitrary disconnections or contamination of water supplies. By contrast, the entitlements include the right to a system of water supply and management that provides equality of opportunity for people to enjoy the right to water.*'[845]

Individuals often organise their own access to water. Especially those living in unserved or underserved areas. For these groups protection against pollution as an element under sustainable development is of great importance. General Comment No. 15 explains, in relation to this, that states should take measures to guarantee that:

> '*Rural and deprived urban areas have access to properly maintained water facilities. Access to traditional water sources in rural areas should be protected from unlawful encroachment and pollution. Deprived urban areas, including informal human settlements, and homeless persons, should have access to properly maintained water facilities. No household should be denied the right to water on the grounds of their housing or land status;*'

Even though traditional water resources are often not secure or safe sources of water, they are still to be protected from pollution and depletion as individuals would be able to use this water, after treatment, for domestic purposes. As mentioned before, principles connected to sustainable development, especially the polluter pays principle and the precautionary principle, could be implemented in order to protect the environment.

843 (Committee on Economic, Social and Cultural Rights, 2002) p. 37.
844 Sancin and Dine. p. 100.
845 (Committee on Economic, Social and Cultural Rights, 2002) par. 10.

5.5.2 PROCEDURAL RIGHTS: ALIGNING THE HUMAN RIGHT TO WATER AND SUSTAINABLE DEVELOPMENT

Aligning sustainable development with international human rights law will allow for both global and regional environmental law complaints procedures and human rights complaints procedures to be invoked when poor environmental protection leads to a violation of human rights such as the right to water.[846] As mentioned in chapter 4 the environmental treaties which outline sustainable development do not include petition procedures for individuals or vulnerable groups. For this reason, human rights tribunals offer the primary opportunity to challenge environmental degradation for which governments are responsible and which have a detrimental effect on the realisation of human rights. Therefore, not only substantive alignment is relevant for the human right to water, but also alignment concerning the procedural aspects.

General Comment No. 15 explains that the human right to water is accompanied by the rights to information, participation and a remedy. Paragraph 48, for instance, explains that:

> *'The right of individuals and groups to participate in decision-making processes that may affect their exercise of the right to water must be an integral part of any policy, programme or strategy concerning water. Individuals and groups should be given full and equal access to information concerning water, water services and the environment, held by public authorities or third parties.'*[847]

These three procedural elements are also mentioned in Principle 10 of the Rio Declaration on Human Rights and the Environment and is central to sustainable development; it states:

> *'Environmental issues are best handled with the participation of all concerned citizens, at the relevant level. At the national level, each individual shall have appropriate access to information concerning the environment that is held by public authorities, including information on hazardous materials and activities in their communities, and the opportunity to participate in decision-making processes. States shall facilitate and encourage public awareness and participation by making information widely available. Effective access to judicial and administrative proceedings, including redress and remedy, shall be provided.'*

The human rights framework and sustainable development are highly complementary in that they both recognise these three procedural elements. The following section discusses the procedural rights which can be found in both environmental as well as human rights law. It explains which complaints procedures and strategies are available and what the limitations are.

---

846 Dinah Shelton, 'The Environmental Jurisprudence of International Human Rights Tribunals' in Romina Picolotti and Jorge Daniel Taillant (eds), *Linking Human Rights and the Environment* (The University of Arizona Press 2003). p. 1.

847 (Committee on Economic, Social and Cultural Rights, 2002) par. 48.

5.5.3 INFORMATION

The right to freedom of information or the corresponding duty for the state to inform are included in many international environmental agreements and human rights treaties. For instance, Article 19(2) ICCPR and Article 19 UDHR include the right to information. However, this right is absent in the ICESCR in which a more elaborate human right to water has been included. However, it is recognised as a general principle of human rights and the CESCR emphasises its place in relation to the human right to water under the requirement of accessibility, stating that: 'accessibility includes the right to seek, receive and impart information concerning water issues.'[848]

5.5.4 PARTICIPATION

As mentioned, the Rio Declaration and the Aarhus Convention have been instrumental in developing participatory tools at the national level. Several international law and policy instruments recognise the importance of public participation in water management such as the 1992 Dublin Statement on Water and Sustainability and the 2000 Hague Declaration. But the Bonn Declaration also states that: 'Water resources management should be based on a participatory approach. Both women and men should be involved and have an equal voice in managing the sustainable use of water resources and sharing of benefits.' The document points out that public participation in water management should take place at the lowest appropriate level.[849]

The participation of the communities and individuals concerned is also guaranteed by international human rights law. This participation should be meaningful and not shallow.[850] States are often deterred from respecting this right as it often requires financial and time investments as well. This is especially so when communities are hard to reach. An aspect that sustainable development offers is the consideration of not only participation by the state and the individuals and communities concerned, but also other relevant stakeholders such as NGOs and private parties.

The right to public participation is highly relevant for vulnerable groups. These communities are the local users of water resources and often lack access to information concerning water planning and are not included in the decision-making processes. Governments can make decisions without including these groups, for instance in relation to the privatisation of water services, large infrastructural enterprises and mining concessions. One of the benefits of including stakeholders is that it allows the inclusion of specific local knowledge about local needs, conditions and processes. This information can inform policy development and contribute to the continuity of

848 (Committee on Economic, Social and Cultural Rights, 2002) par.12.

849 ???

850 WaterAid, 'Rights-Based Approaches to Increasing Access to Water and Sanitation' (2011). p. 44; Anne Hellum, Patricia Kameri-Mbote and Barbara Van Koppen, *Water Is Life* (2015). p. 64.

interventions. The participation of stakeholders also enhances the legitimacy of the outcome and the process and improves the sense of ownership of the outcome by the stakeholders.

States should therefore include meaningful participation in the decision-making processes concerning water. An example of good practice in this regard can be found in the safe water communities in Nicaragua.[851] Here the local community has participated in the design, construction and management of the implemented systems. As a result, the community has developed a greater willingness to contribute financially to maintain the systems and also has gained an understanding about the importance of safe water.

5.5.5 REMEDY

A human right will be ineffective if it does not provide right-holders with a remedy provided by the duty bearer when the right is infringed. This is also true for the human right to water. In line with international law in general and international human rights law specifically[852], a state must provide for a remedy within its legal system which, in the event of a violation, serves many purposes such as:

> *'to provide injunctive relief to stop harmful environmental activities, to award damages to compensate for the harm suffered, restitution or remedial measures; and to award litigation costs and fees.' Remedies can thereby serve as a preventative measure and could be used for redressing actual violations of an individual's right. Having an available remedy is in itself a right which provides redress in the event of violations of rights and deters infringements from taking place.'*

And again this requirement can bridge international human rights law and sustainable development. General Comment No. 15 explains:

> *'Any persons or groups who have been denied their right to water should have access to effective judicial or other appropriate remedies at both national and international levels (see General Comment No. 9 (1998), para. 4, and Principle 10 of the Rio Declaration on Environment and Development). The Committee notes that the right has been constitutionally entrenched by a number of States and has been subject to litigation before national courts. All victims of violations of the right to water should be entitled to adequate reparation, including restitution, compensation, satisfaction or guarantees of non-repetition. National ombudsmen, human rights commissions, and similar institutions should be permitted to address violations of the right.'*[853]

---

851 De Albuquerque, 'Report of the Special Rapporteur on the Human Right to Safe Drinking Water and Sanitation'. par. 76.

852 See (Committee on Economic, Social and Cultural Rights, 1990) paragraph 5 Bulto, *The Extraterritorial Application of the Human Right to Water in Africa*. p. 227.

853 (Committee on Economic, Social and Cultural Rights, 2002) par. 55.

The right to a remedy entails both procedural as well as substantive aspects. The procedural aspect guarantees that individuals have access to an impartial and competent tribunal, while the substantive aspect ensures the right to receive relief in case of an infringement or the threat thereof. This can be provided in the form of restitution, compensation or other means.[854]

Sustainable development provides for principles which can provide guidelines for states in providing remedies, such as the polluter pays principle. In *Vellore Citizens Welfare Forum v. Union of India* this approach was followed. The Vellore Citizens Welfare Forum was concerned about the 900 tanneries in the State of Tamil Nadu which caused water pollution affecting the Palar River which is the main source of drinking water in the area and for irrigating agricultural land. In this case the Court recognised the economic importance of these tanneries. However, it argued that their activities should not destroy the environment and referred to the concept of sustainable development. In line with this reasoning the Court ordered the government to implement the polluter pays and precautionary principles and to provide compensation for those who had been negatively affected by the pollution.[855] The Court also explain that: '*The "Polluter Pays" principle as interpreted by this Court means that the absolute liability for harm to the environment extends not only to compensate the victims of pollution but also the cost of restoring the environmental degradation.*'[856] This a prime example of how the polluter pays principle can be implemented in such a way that it will deter third parties and the government from polluting water and also how the income gained through instruments such as fines can be earmarked to contribute to resolving the issue. In this sense, sustainable development greatly supports the human right to water.

However, in the event that remedies at the national level have been exhausted or are non-existent, remedies at the international level could provide an appropriate option. Individuals and groups who have been deprived of an opportunity to obtain redress at the national level may bring an action before an international court of an international non-judicial body such as the CESCR. However, in order to lodge a complaint with the Committees which oversee human rights covenants, such as the CESCR, the perpetrator state must be a party to the relevant treaty and have accepted the competence of the relevant Committee to examine individual complaints. Even though the Committee's decisions are not legally binding, they are considered to be authoritative and states are encouraged to give effect to its findings and recommendations. Also, the interaction between the Committee, the complainant and the state can have a positive effect on the right to information, as states are invited to provide information, which could assist the complainant and other individuals to facilitate their access to water. In addition, individual complaints allow for the rights mentioned in treaties to shape their concrete

854 Shelton, 'The Environmental Jurisprudence of International Human Rights Tribunals'. p. 8.

855 Winkler, 'Judicial Enforcement of the Human Right to Water – Case Law from South Africa, Argentina and India'. p. 13.

856 Vellore Citizens Welfare Forum vs Union Of India & Ors on 28 August, 1996 https://indiankanoon.org/doc/1934103/

meaning which is highly relevant for the human right to water as it is a right which is still developing. The entry into force of the Optional Protocol to the CESCR means that the right can be further elaborated and will demonstrate its function in the specific context of the state parties.[857]

Another situation which requires effective remedies is where individuals need remedies for addressing transboundary violations of the human right to water. As explained above, pollution and the depletion of shared water resources may affect access to safe drinking water for inhabitants of co-riparian states. However, effective measures of redress are currently lacking in this area for individuals and groups as most instruments concerning shared water resources focus on interstate relations. Including individuals and groups could positively influence sustainable water management and the inclusion of vulnerable groups in the process. If individuals and groups would not be dependent on their state to initiate a claim against a foreign perpetrator state, these states would not assume impunity in relation to their human rights violations.[858] It is often the case that states do not consider bringing actions against co-riparian states because of economic or political considerations, leaving infringements unresolved. In this respect, allowing individuals and groups to bring an action independently will encourage states to respect their duty to protect by holding third parties located within their borders accountable for human rights infringements.[859] And lastly, the direct access of individuals to 'foreign judicial and other institutional avenues to bring the acts and omissions of riparian states under proper scrutiny means that states would now be forced to take people's view into consideration before or in the course of the allocation and use of shared water resources.'[860]

5.6 CONCLUSION

The above has demonstrated which elements from sustainable development can support the implementation of the human right to water for vulnerable individuals and groups. This chapter thereby provides an answer to the main research question of this study: How can the concept of sustainable development contribute to the sustainable realisation of the human right to water for vulnerable groups? This study has chosen to do this by aligning the two concepts and using elements of sustainable development to elaborate the requirements of the human right to water. Discussing the human right to water from an environmental perspective broadens our understanding of human rights, their content and their effective implementation. Sustainable development is appropriate for complementing the right as it sets standards for how the resource on which the right depends should be used. This study explores those soft and hard law

---

857 See Catarina de Albuquerque.

858 Bulto, *The Extraterritorial Application of the Human Right to Water in Africa*. p. 233.

859 (Bulto, 2014) p. 233.

860 (Bulto, 2014) p. 233.

requirements which deal with the protection and conservation of water resources and which themselves form the outer boundaries of the human right to water. They are the requirements which facilitate the effective implementation of this right even if they are not direct obligations under the right.

Sustainable development and human rights are compatible as they are human-centred and take aspects such as the economy and social development into consideration. However, they differ in their scope as well as the nature and target of their obligations. The human right to water allows for individuals to take their concerns to court while sustainable development has mostly been an interstate affair.

The first necessary step for realising a sustainable human right to water for vulnerable groups is the prioritisation of water for domestic purposes. It was established that the human right to water sets such a priority and that additional lines of reasoning can be found in international water law. This prioritisation will push states to reallocate water resources from other water uses so that these resources can then be used for domestic purposes.

When integrating sustainable development, the general obligations of progressive realisation and using the maximum available resources provide a welcoming place for concepts such as equitable use and limiting uses which harm ecosystems. Besides the adjustments that sustainable development can make to the human right to water, other considerations such as the types of water resources should be used, and understanding the way in which vulnerable groups use water should play a role as was shown in chapter 4.

In addition, international human rights law is still ineffective in the area of extraterritorial obligations which would allow for individuals in one state to bring a claim against the government of another state.[861] Extraterritorial obligations would provide redress in those situations where states externalise pollution or engage in unsustainable use patterns at the expense of individuals in other states.[862] The entry into force of the Optional Protocol to the Covenant on Economic, Social and Cultural Rights may provide a window of opportunity in this respect.[863] This, in turn, will assist in creating 'a reliable and effective system of environmental protection which would help ensure the well-being of future generations as well as the survival of those persons, often including indigenous or economically marginalised groups, who depend immediately upon natural resources for their livelihoods.'[864]

861 McCaffrey. p. 20 Rijksuniversiteit Limburg Centre of Human Rights., International Commission of Jurists (1952- and Urban Morgan Institute for Human Rights., 'The Maastricht Guidelines on Violations of Economic, Social and Cultural Rights' (1998) 20 Human Rights Quarterly 691 <http://muse.jhu.edu/journals/human_rights_quarterly/v020/20.3intl_comm.html> accessed 4 February 2016.

862 McCaffrey. p. 20.

863 See Catarina de Albuquerque.

864 A. Boyle, M. Anderson (eds), Human Rights Approaches to Environmental Protection (Oxford: Oxford University Press, 1996).

CHAPTER 6

TEN BUILDING BLOCKS FOR A SUSTAINABLE HUMAN RIGHT TO WATER

An Integrated Method to Assess Vulnerable Groups' Sustainable Access to Water for Domestic Purposes

'The lesson to be learnt is that access to water depends on a range of economic, political, technical, cultural, financial, organisational, social and managerial issues. Because of the multi-faceted nature of the problem a holistic view on water service provision is imperative.'[865]

6.1 INTRODUCTION

In the quest to realise the human right to water for all, states are experiencing many challenges, especially concerning the fulfilment of their obligations towards unserved and underserved populations. With natural resources becoming more and more scarce, states with limited capacities are increasingly required to review their overall water management in relation to providing access to water for domestic purposes. The human right to water can play a crucial role in assisting the development of a context-specific policy and regulations. Placing the human right to water within a specific national context will put it in a particular physical context regarding water resources. This right, however, is still developing and its position within the overall system of water management is unclear. This book therefore provides a method by which the human right to water can be observed within the bigger picture of sustainable water management.[866] As such this chapter answers the question: How can a sustainable human right to water be assessed within the local context?

---

865 Björklund and Sjödin. p. 7.

866 A short version of the method has been published in: Daphina Misiedjan, 'Ten Building Blocks for a Sustainable Human Right to Water: An Integrated Method to Assess Sustainable Access to Water for Domestic Purposes of Vulnerable Groups' in Erin Daly and others (eds), *New Frontiers in Environmental Constitutionalism* (UNEP 2017).

This chapter presents an assessment method in which the rights-based approach to access to water and relevant environmental principles have been integrated with the aim of facilitating the sustainable enjoyment of the right to water by vulnerable groups. This assessment method will be referred to as the 'Ten building blocks for a sustainable human right to water' and was built out of a multi-disciplinary assessment method for water governance developed by Van Rijswick, Edelenbos, Hellegers, Kok, and Kuks (2014). The method as developed by Van Rijswick et Al. is a three-step multi-disciplinary method to assess how states tackle different water issues such as water shortage, water quality and flood risks.

This method can be used to pose specific questions in relation to water management. It has therefore been adjusted in order to assess relevant aspects for the implementation of a sustainable human right to water for vulnerable groups and thereby extends the applicability of this approach. It brings together the different disciplines of law, economics, and public administration and provides for basic water system knowledge, which is needed for a comprehensive exploration of the main challenges and for offering recommendations which are specific to a national or regional context. The method which has been developed is of a diagnostic nature.[867] It serves to identify the strengths and weaknesses in water governance, with the aim of providing recommendations for dealing with water issues in an efficient and sustainable manner. For this reason, the method can be directly applied in national contexts. This assessment method has been used to assess the realisation of the human right to water in Yemen. And the next chapter includes its application in two case studies in Suriname.

6.2 INTRODUCTION TO THE ASSESSMENT METHOD

The following section introduces the assessment method. The previous chapters have demonstrated that in order for the human right to water to be realised, several elements are important to analyse. These conclusions are summarised in this section. The conclusions of the previous chapters have been adjusted to serve as elements in the assessment framework. This chapter adopts a water governance approach due to the complex nature of water systems and the decisions made concerning these systems. It is difficult to formulate an all-encompassing definition of water governance. The Global Water Partnership explains that 'water governance relates to the range of political, social, economic and administrative systems that are in place to develop and manage water resources and the delivery of water services at different levels of society.'[868] Governance looks at all relevant parties and not merely the government. Even though, according to the human rights approach, the government has the final responsibility, other parties also contribute to or diminish efforts towards the realisation of the human right to water for vulnerable groups and are relevant to be discussed within this assessment method.

867 Van Rijswick and others. p. 2.
868 Charles Batchelor p. 1.

Sustainability as a criterion for the realisation of the human right to water

The Brundtland Report formulated the classic definition of sustainable development as 'development which meets the needs of the present without compromising the ability of future generations to meet their own needs.'[869] The principles of this definition of sustainable development were formulated at the United Nations Conference on Environment and Development in 1992. Sustainability has relevance for water management in general[870] and the human right to water in particular. The (former) Special Rapporteur on the Human Right to Water considers sustainability to be a human rights principle and therefore a cross-cutting criterion.[871] Sustainability has special relevance for realising the human right to water, especially in relation to the human rights core concepts of progressive realisation and non-retrogression.[872]

What this entails is that water must be provided in a manner that respects the natural environment.[873] But also that consideration needs to be given to the different dimensions of economic, social and environmental sustainability.[874] Furthermore, services and facilities must be available for present and future generations and should not hinder the ability of future generations to meet their needs. Sustainability requires that in using water resources for realising the human right to water both the protection of water quality as well as implementing measures to deal with droughts and other water issues are taken into consideration. Therefore, the sustainable use of water resources requires a holistic approach to water management.[875] This adjusted assessment method places sustainability in the human rights framework. It uses environmental principles which have been connected to sustainable development and specifies implementation to establish indicators of compliance.

Water and vulnerable groups

Vulnerable groups have been targeted for a number of reasons. Firstly, because those who currently lack access to water are predominantly vulnerable groups. Due to the Millennium Development Goals the global community at large and states individually made efforts to increase access and aim for universal access to water. Different targets were reached. However, it seems that those targets were what can be described as 'low hanging fruit'. This means that those which were easiest to reach were targeted in order to attain the required percentages. As a result, the most disadvantaged and vulnerable

---

869 Brundtland Commission.

870 See Dellapenna, 'Is Sustainable Development a Serviceable Legal Standard in the Management of Water?'

871 Albuquerque.

872 De Albuquerque, 'Report of the Special Rapporteur on the Human Right to Safe Drinking Water and Sanitation'. p. 1.

873 Committee on Economic Social and Cultural Rights. paragraph 11.

874 De Albuquerque and Roaf, 'Principles'. p. 75.

875 Dellapenna, 'Is Sustainable Development a Serviceable Legal Standard in the Management of Water?' p. 89.

groups have missed out on these developments.[876] Consequently, when discussing improving access to water and realising the right to water universally, it is important to recognise that those who currently lack access to water are predominantly vulnerable and disadvantaged groups, those who are not easily reached and are mostly dependent on traditional sources of water. States have the immediate obligation to adopt measures against inequalities in access and prioritising access for vulnerable groups should be a focal point in the policies and practices of states.[877] Secondly, the situation of vulnerable groups serves as a litmus test for the efficacy of the human right to water in a state. And, thirdly, by 2030 the world is expected to have a 40% shortfall in water supplies if the current trend continues. This means that the number of people lacking access to water will increase. If current water management will adapt to cover access to water for vulnerable groups, the increase in vulnerable groups will be more effectively dealt with in the future as was explained in chapter 2.

Identifying vulnerable groups

The assessment method looks specifically at the situation of vulnerable groups.[878] The term vulnerable groups in this context refers to those groups who are susceptible to not being able to secure sustainable access to adequate quantities of water of an acceptable quality for domestic purposes due to social and/or environmental factors.[879] This definition expands on the concept of vulnerability developed by the UN Committee on ESCR.

The UNCESCR does not currently make use of a specific definition of vulnerability. Neither is there a list of such groups or any characteristics which have been agreed upon.[880] However, in a few General Comments specific groups have been identified as being at risk in relation to a particular right.[881] The Committee has also done this in relation to the human right to water in General Comment No. 15. Here it mentions that specific groups 'have traditionally faced difficulties in exercising this right'.[882] Among these groups are women, children, refugees and migrant workers. These groups can be seen as vulnerable in a specific situation, and the Committee encourages states to take specific measures to target these groups. What can be observed is that the Committee

876 JMP, World Health Organization and UNICEF, 'Ending Inequalities – a Cornerstone of the Post-2015 Development Agenda' 1.

877 De Albuquerque and Roaf, 'Principles'. p. 15.

878 The term vulnerable has been chosen as opposed to disadvantaged or marginalized, due to its cross-disciplinary use. Similar to the issue of water access, the term vulnerability appears to be relevant within the intersection between environmental science and human rights.

879 Developed from the term 'water security' (UN-Water, 2013). Water security is defined as the capacity of a population to safeguard sustainable access to adequate quantities of acceptable quality water for sustaining livelihoods, human well-being, and socio-economic development, for ensuring protection against water-borne pollution and water-related disasters, and for preserving ecosystems in a climate of peace and political stability.

880 (Chapman & Carbonetti, 2011) p. 683.

881 Chapman and Carbonetti. p. 724.

882 Committee on Economic, Social and Cultural Rights. par. 16.

mostly takes into account social factors and does not explore the relationship between social and environmental factors which render certain groups and individuals vulnerable in relation to accessing water for domestic purposes.

The Committee takes a narrow view concerning geographical and environmental aspects leading to vulnerability. In General Comment No. 15 it mentions that rural and deprived areas should have access to proper water facilities. These areas include, for instance, informal human settlements. It also demands that states protect water resources from contamination.[883] However, it does not recognise environmental factors as a characteristic of vulnerability which can also be termed 'environmental vulnerability'. Environmental vulnerability is used to describe the degree to which individuals and groups are susceptible to environmental damage.[884] The reason for supplementing the categories of vulnerable groups provided by the CESCR is that the environment itself can be another reason for vulnerability and can actually exacerbate other forms of vulnerability. These categories of vulnerability together determine the obstacles which individuals and groups face in realising their human right to water.[885]

There are several benefits in including environmental vulnerability when assessing the realisation of the human right to water. The benefit of including environmental vulnerability in the human rights framework is that those groups who are disproportionately affected by environmental degradation can then also be the recipients of targeted measures. Also, the challenge of vulnerability is that it is not necessarily concerned with the physical scarcity of resources but with access to those resources.[886] Therefore environmental vulnerability should be considered in relation to social vulnerability as discussed by the UN Committee on CESCR. This is because even in situations where water is abundant, certain groups and individuals are still vulnerable to having their access to water restricted, regardless of the availability of water.

Vulnerable groups should be identified in the national context by taking into account the groups suggested by the UN Committee on CESCR. The categories provided by the Committee do not apply in and of themselves. This assessment method identifies not only those which have been mentioned in General Comment No. 15 but also looks directly at traditional water sources and how the state of these resources affects those who directly depend on them. For instance, if a village depends on a nearby river for its access to water and this river becomes contaminated leaving the water unsuitable for domestic purposes, this village then becomes vulnerable to not being able to secure access to water for its inhabitants. What is to be taken into account is the groups' strategies to adapt especially in those circumstances where infrastructure is lacking. Villagers may for example buy water or harvest rain water.

---

883 Ibid. par. 8.
884 Barnett. p. 257.
885 See also chapter 3.
886 Hermans. p. 23.

6.3 THE BUILDING BLOCKS OF THE ASSESSMENT METHOD

The previous sections summarised the conclusions of the previous chapters. From these conclusions criteria are distilled which are included in the assessment method. The following sections explain the ten building blocks for the assessment of sustainable access to water for vulnerable groups as illustrated by Figure 6.1. Each block has its own criteria which are leading concepts in the analysis of national contexts.

Figure 6.1 Assessment of sustainable access to water for vulnerable groups

Each of the ten building blocks is described in the following sections.

1. Water system knowledge

Studying the knowledge which relevant actors have about the water system is essential for realising the human right to water. It can enhance the sustainability of this right for vulnerable groups as actors will understand how much water is available and how it should be managed. The water system is defined by Rijswick et al. 2014 as:

> *'the combination of natural physical resources (such as rivers, rainfall, seas, lakes etc.) and man-made infrastructure (such as canals, pumping stations, reservoirs, flood defences etc.). The system supports societal functions, such as domestic and industrial water use, irrigation, shipping, hydropower, safety, etc., and includes the ecosystems related to water.'*[887]

887 Van Rijswick and others p. 4.

A distinction is made between natural physical resources and man-made infrastructure. The focus within this building block is on the sustainability of the water system that supports domestic water use, which is defined by the UN Committee on ESCR as follows: drinking, personal sanitation, washing of clothes, food preparation, personal and household hygiene.[888] Therefore the water resources and infrastructure which directly and indirectly support domestic uses should be studied.

The human right to water sets out certain recommendations and requirements for the infrastructure to realise this right. However, in relation to the water resources themselves it only determines a limited number of standards. For instance, it discusses quality and quantity standards at a service point and these standards are not yet connected to sustainability requirements. General Comment No. 15[889] does mention that the realisation of the right must be sustainable but does not elaborate on the way in which this should take place. For this reason, the information collected in this building block assists in this respect. Sustainability in relation to water system knowledge "means that water quality and availability have to be ensured in a way that respects and supports the larger environment. Water contamination and over-extraction must be avoided in order to ensure continued access to safe and sufficient water."[890] As a result states must determine what uses are sustainable on the basis of a very fact-specific analysis of the appropriated uses of a particular water resource in a particular context.[891]

For instance, states could take sustainability into account while deciding which water resource is appropriate to be used for drinking purposes. The water source must provide clean drinking water but also have sufficient water available. Here a state should take the hydrological cycle into account, especially as some resources are considered to be finite such as certain groundwater resources.[892] The hydrological cycle is very much influenced by human activity, as humans collect rain and extract great amounts of groundwater and surface water from lakes and rivers in order to meet the demand by the agricultural sector, households, industry and the energy sector. Human activity does not merely influence the quantity of water which is available but also its quality.[893] In addition, climate change also influences this cycle. As a consequence, water resources

---

888 Committee on Economic, Social and Cultural Rights. par. 12 "In this context, "drinking" means water for consumption through beverages and foodstuffs. "Personal sanitation" means disposal of human excreta. Water is necessary for personal sanitation where water-based means are adopted. "Food preparation" includes food hygiene and preparation of food stuffs, whether water is incorporated into, or comes into contact with, food. "Personal and household hygiene" means personal cleanliness and hygiene of the household environment."

889 Ibid. par. 7.

890 De Albuquerque, 'Report of the Independent Expert on the Issue of Human Rights Obligations Related to Access to Safe Drinking Water and Sanitation "Compilation of Good Practices"'. p. 16.

891 Dellapenna, 'Is Sustainable Development a Serviceable Legal Standard in the Management of Water?' p. 89.

892 See for instance: Ottmar Edenhofer and others, *Climate Change, Justice and Sustainability: Linking Climate and Development Policy* (Springer 2012) <http://link.springer.com.ezproxy.usherbrooke.ca/chapter/10.1007/978-94-007-4540-7_5>.

893 World Health Organization, 'Guidelines for Drinking-Water Quality' (2011).

are currently over-allocated and heavily polluted meaning that more assignments have been made than can be sustained.[894]

Considering these aspects ties into other characteristics such as affordability which is highly important for vulnerable groups such as the poor. Choosing water from a source which needs significant treatment or has limited water available raises the costs which can lead to increased prices for water supply services. When assigning different water uses to different resources, states should be aware of the social functions that these water resources have. They should for example question how many individuals and households rely on certain water resources such as rivers or creeks. Especially vulnerable groups live in areas which are currently unserved.[895] As a result they rely on traditional unimproved water sources for the realisation of their human right to water.

What is often omitted from the water uses equation are the environmental functions of water. As far as the sustainability of water resources is concerned, states should also ask how much water ecosystems actually need. If ecosystems are not maintained, future generations will be deprived of access to water. Knowledge of the water system should therefore also include information to support sustainable water use which does not undermine 'the integrity of the hydrological cycle or the ecological systems that depend on it.'[896]

Assessment criteria: Is there sufficient knowledge of the existing water system in order to support access to water for vulnerable groups? If not, what are the gaps; is sufficient knowledge available to assess the impact on the water system because of changes in environment and societal functions?[897]

2. Responsibility, authority and means

The means question whether the tasks and competences assigned have been done so at the right level of the organisation or government. Decentralisation is often sought in order to enhance participation and to adopt subsidiarity.[898] Capacity also deals with whether the right agency has been assigned and whether cross-sectoral coordination can take place with regard to financial, natural, human, organisational, technological, information and other resources.[899]

Property rights and water rights

The ownership of the water supply infrastructure or the right to use water indicates power and control. The role and responsibilities that come along with these rights relate

---

894 (van Rijswick et al., 2014) p. 4.

895 UNICEF and WHO. p. 23.

896 Peter H Gleick, *California Water 2020: A Sustainable Vision* (Pacific Institute for Studies in Development, Environment, and Security 1995). p. 100.

897 Van Rijswick and others. p. 5.

898 Ibid. p. 10.

899 Categories as suggested by Robertson.

to water governance and management in a particular manner and are rooted in formal and informal legislation. Central issues are often how property rights are defined, who benefits from these rights and how the enforcement of these rights is organised. With changing patterns in water supply and demand, amendments to existing legal frameworks are often required. What is needed are well-defined, coherent rules, roles and responsibilities through legislation of formal and informal water rights. This is assumed to lead to both social and economic advantages as well as environmental benefits.[900]

For instance, when individuals who were previously denied access to water rights are given such a right or are recognised to have such a right, they often assume ownership. This means that they feel responsible for how the water resource is being used, managed and dealt with. They thereby have an incentive to manage the source sustainably.

Responsibility

States are the main duty bearers under international human rights law. Therefore, the state bears primary responsibility for fulfilling the human right to water which then creates certain obligations. However, states can delegate tasks to third parties. As explained, the obligations of the state can be divided into three categories. Firstly, the tripartite typology also known as the responsibilities to respect, protect and fulfil. Secondly, the general obligation to realise the right to water progressively and, lastly, the minimum core obligations. The following section delineates the obligations of states and sets out the criteria by which to judge the strength of this building block.

Obligation to respect

The obligation to respect requires states not to interfere in the enjoyment of human rights. Therefore, states are required not to take any measures that would undermine individuals' ability to have access to water. More specifically, states must respect any existing water supply in line with human rights standards.[901] To further explain this obligation several examples can be found in chapter 3.

Individuals as rights holders are expected to use the options which are available to them for realising the right to water. As a consequence, individuals look for the most convenient possibilities to meet their needs. They for instance use water from resources near their dwellings such as creeks, wells and rivers for domestic purposes. These sources are often considered to be unsafe and for this reason they are contrary to the standard that the human right to water sets. Nevertheless, certain nuances can be applied. Firstly, these water resources can become appropriate for domestic use when they are treated. Secondly, it can be argued that even though water from these sources is not suitable for drinking, it can still be of sufficient quality for other purposes such as washing items of clothing or sanitation. Thirdly, it is often the only water which is

900 Charles Batchelor p. 8.

901 Winkler, *The Human Right to Water Significance, Legal Status and Implications for Water Allocation*. p. 108.

available to them. Therefore, even though the water for unsafe sources does not meet human rights standards, states are required to respect these as an existing water supply and therefore refrain from polluting these sources or limiting access to them.

Secondly, the obligation to respect also concerns water allocation. When sufficient resources have been assigned for meeting basic needs, states must respect this allocation and not reallocate the resources for other uses such are agriculture or hydro-electricity.[902]

Thirdly, regarding water supply services states are prohibited from arbitrarily disconnecting water services for those who can no longer afford them. This is because no one should be deprived of having access to a minimum essential level of water. As a consequence, states should take appropriate measures to facilitate continuous access for those who lack the means to secure this access.

Therefore, in order to meet the obligation to respect, states need information about how dependent individuals are on water resources. If states are aware of certain water resources being used by individuals and communities, hindering their access to this water by either polluting the source or physically obstructing access by these individuals can be seen as undermining their obligation to respect. In this regard, the example of the San living in the Kalahari Game Reserve can be recalled.[903] Even though the quality of water available from the well does not meet human rights standards, as it requires treatment in order to make it appropriate for drinking, access to that well should still be respected. This is because individuals will be able to use this water to meet their needs and it is the only water which is readily available to them.

Obligation to protect

The obligation to protect requires states to 'prevent third parties from interfering in any way with the enjoyment of the right to water.'[904] Third parties consist of individuals, groups, corporations and other entities. This obligation is especially important when the water services are operated and controlled by a third party. In this case, the state is required to adopt an effective regulatory system as a way of keeping an eye of the service provider. Also, states must protect their citizens from water pollution caused by others. A measure which may be implemented can be based on the Polluter Pays Principle and this would complement the realisation of the human right to water.

Obligation to fulfil

States are required to facilitate, promote and provide for the rights of individuals and other positive obligations that are contributory. The state needs to take an active role in the realisation of the right to water for individuals and this can be through legislation, the judiciary, the administration and other measures. It also means that the state is

902 Ibid. p. 108.

903 James R May and Erin Daly, 'Water and Environmental Constitutionalism', *Global Environmental Change* (Cambridge University Press 2015). 193.

904 Committee on Economic, Social and Cultural Rights.

obliged to provide what is necessary for the enjoyment of a right if groups or individuals are unable to provide for it themselves. For instance, in the case of the human right to water states are obliged to make water accessible at reasonable prices for those who cannot afford it or even for free if necessary. Connected to this obligation is the requirement for states to use the maximum available resources with the objective of the progressive realisation of the human right to water.

Means and capacity

Regarding the use of means, the human right to water sets out certain requirements that states should take into account. The requirements for the use of resources are related to the level of the intended services. Or to put it in human rights terms, the requirements deal with the immediate realisation of the minimum core obligations and that of the progressive realisation of the human right to water. When it concerns minimum core obligations, such as access to drinking water to sustain life, states are expected to be able to meet the requirements and to use all necessary resources as a matter of priority to bring this about.[905] Beyond the minimum core, states are expected to make use of the maximum available resources to progressively realise the full enjoyment of the human right to water.

As mentioned in chapter 3 in the context of human rights, resources have mostly been explained as meaning financial resources. In the situation of the human right to water, water resources and their allocation are highly relevant. Therefore, the understanding of resources as meaning all relevant resources needed for the realisation of human rights will be followed. This includes: financial, natural, human, organisational, technological, information and other resources.[906]

In addition to the means, sufficient capacity needs to be available for realising the human right to water in a sustainable manner. Developing countries often lack the capacity to promote more sustainable use patterns.

Assessment criteria: Which parties have which responsibilities in realizing the human right to water for vulnerable people? Are the responsibilities clear and are their responding mandates also clear? Are contracts between States and service providers fully compliant with human rights and sustainability standards? Are there capacity-building strategies for stakeholders?

3. Stakeholder participation

In the provision of a water supply for domestic purposes and connected water management different parties are involved. Governments have final responsibility for providing access to water for domestic purposes for those within their territory,

---

905 Winkler, *The Human Right to Water Significance, Legal Status and Implications for Water Allocation.* p. 122.

906 Categories as suggested by Robertson.

including vulnerable groups and individuals. Nonetheless, governments are often dependent on many different stakeholders for its realisation. 'In this context, governments sometimes deliberately, sometimes forced by circumstances, give more room to *certain* stakeholders to influence decision-making and implementation which often have different values and interests.'[907] How these different interests and values influence access to water for vulnerable groups is highly relevant when assessing the sustainability and the general exercise of this right. This process of influencing and balancing is covered within this building block of stakeholder participation.

The principle of participation is enshrined in many human rights conventions as well as environmental conventions. In this sense, the practice and theory surrounding this principle within these two fields will enhance the application thereof regarding the implementation of a sustainable human right to water. Participation in the human rights sense qualifies as 'active, free and meaningful participation.'[908] The principle from the environmental perspective, on the other hand, includes different aspects, determining that participation includes three aspects: access to information, public participation in decision-making and access to justice. These aspects are considered to be the clearest elaboration of Rio Principle 10 as it is presented in the Aarhus Convention.[909] Even though these aspects are recognised within the human rights field they are often discussed as separate general principles[910] or in relation to specific groups such as women or children. The Human Rights Council, for instance, calls for the full, free and meaningful participation of local communities.[911] What incorporating Principle 10 contributes is a specific environmental point of view so that whoever is affected has a means of redress, which is highly relevant for the right to water. Especially under those circumstances where environmental issues affect the quality and quantity of the drinking water available to vulnerable groups.

From a human rights perspective, several levels of participation need to be assessed. Both the process of decision-making as well as the outcome are relevant for assessing whether stakeholders are able to meaningfully participate. The Human Rights Council has explained that this also applies to the three phases of water management strategy: design, implementation and monitoring.[912]

This framework discusses the participation of all relevant stakeholders in this building block and not merely that of vulnerable groups. This is because what is important is assessing how the different viewpoints are presented, represented and weighed. Other stakeholders which may not be directly involved in the water supply

907 Van Rijswick and others. p. 7.

908 General Assembly, Declaration on the Right to Development, 4 December 1986, A/Res/41/ 128, Art 2(3).

909 ECE Convention on Access to Information, Public Participation in Decision-making and Access to Justice in Environmental Matters, Aarhus, 25 June 1998, entry into force: 30 October 2001.

910 For instance, the Convention on the Rights of the Child assures that children should be able to speak their mind freely in matters that affect them and that their views should be taken into account.

911 Human Rights Council resolution 12/8, para. 4.

912 Human Rights Council resolution 12/8, para. 4.

for domestic purposes for vulnerable groups may nevertheless affect their water access. Consequently, the relationship between, for instance, water suppliers and rights holders but also the relationship between different rights holders amongst themselves will be assessed.

In the decision-making processes that may affect the exercise of the right to water for individuals and groups, their right to participation must be respected and be considered to be an integral part of any policy, programme or strategy relating to water.[913] The Human Rights Council calls for the full, free and meaningful participation of local communities in the following three phases of these strategies: design, implementation and monitoring.[914]

Assessment criteria: Are vulnerable groups identified and involved in the water management which affects them? Are vulnerable groups freely, actively and meaningfully involved? How are the interests, concerns and values balanced in the problem analysis and the decision-making concerning the design, implementation and monitoring of the solution?

4. Values, principles and policy discourses

Values

Values are believed to be important in a society. These values control behaviour and assist in making decisions. In relation to water management, values help in prioritising water allocation. Values can be elaborated as principles and translated into policy and regulation.[915] What should be kept in mind is that multiple values can be upheld in society. How these values are followed up by principles, policy and legislation determines the hierarchy of values. Examples of commonly known values are: justice, respect and dignity.[916] It is often argued that societies across the globe have a common set of values.[917] However, how these values are interpreted may differ between societies. These common values have contributed to the development of universal human rights.

Principles

General human rights principles apply to the human right to water. These principles are non-discrimination, equality, access to information, the right to participate, sustainability and non-retrogression. Along with the human rights principles a selection has been made of environmental principles as mentioned in chapter 4 which can be integrated into national strategies in order to realise the human right to water for vulnerable groups. Environmental legislation based on these principles

---

913 Committee on Economic Social and Cultural Rights. p. 15.
914 (De Albuquerque, 2010) p. 8.
915 Van Rijswick and others. p. 5.
916 (van Rijswick et al., 2014) p. 5.
917 Https://documents-dds-ny.un.org/doc/RESOLUTION/GEN/G12/173/96/PDF/G1217396.pdf?OpenElement.

can complement the human rights framework.[918] These principles are articulated in numerous treaties concerning environmental law including the Rio Declaration on the Environment and Development. The following principles have been found to be the most relevant for this framework: intergenerational equity, integration, common but differentiated responsibilities, public participation, precaution, the polluter pays principle, environmental impact assessment and international cooperation.

The assessment framework studies which principles have been translated into national policies and legislation by states and how possible tensions have been resolved. For instance, the translation of the polluter pays principle can be effective in protecting water resources used by vulnerable groups from pollution.

Assessment criteria: Is there sufficient knowledge of shared or conflicting values, viewpoints and principles (represented by different policy discourse coalitions) for water issues and their consequences for facing water management issues?

5. Regulations and agreements

Appropriateness

Whether the regulations and agreements are suitable within the context in which they have to function depends on a number circumstances. The circumstances which this assessment method looks at are the cultural, political, institutional, legal and economic circumstances. For instance, the legal traditions and which legal systems are present within a nation will influence whether the agreements and regulations are appropriate. If, for example, a national legal context is pluralistic, regulations and agreements will need to be harmonised and/or be made to reinforce one another.[919] Otherwise the regulatory system will be subject to internal conflicts. Also, agreements and regulations need to fit within the context in which they function. Often regulations which have been effective in one context are transplanted into another legal system and context. However, this exercise may not yield the same results as such regulations may not translate well into the other circumstances or contexts. Also, the regulations should be able to function in the physical context where they need to function.[920]

Legitimacy

Agreements and regulations need to be considered as legitimate sources of law. The legitimacy of regulations and agreements is expected to enhance the acceptance of norms by stakeholders.

---

918 Tully. p. 57.

919 Martín Iribarnegaray and Lucas Seghezzo, 'Governance, Sustainability and Decision Making in Water and Sanitation Management Systems' (2012) 4 Sustainability 2922 <www.mdpi.com/2071-1050/4/11/2922/> accessed 26 February 2014. p. 2926.

920 See for instance the example of mobile water courts in THIGL.

Legal certainty and adaptiveness
Norms and rules should strike a perfect balance between adaptiveness and legal certainty.[921] The adaptiveness of norms can be realised by including open norms and having a procedure which allow for flexibility. Striving for sustainability in the realisation of the human right to water requires adaptiveness on the part of the regulatory framework. The availability of water may fluctuate due to, for instance, climate change and pollution and the norms must be able to adapt to these changes. Nevertheless, the regulatory system is also in need of enhancing legal certainty. Stakeholders need to be able to anticipate and be able to base their behaviour on the rules that they are aware of.

Assessment criteria: How do regulations and agreements recognise or support a sustainable human right to water? What are the main problems with regard to the above-mentioned aspects?

6. Trade-offs between social objectives: service-level agreements

Realising sustainable access to water for vulnerable groups is not solely an issue of establishing the appropriate rules and norms and implementing them as will be explained in subsequent building blocks. Realising sustainable access is to be seen as a continuing process of negotiating and trade-offs between the interests of stakeholders.[922] This ties in with the governance perspective which takes the different roles, authorities and parties into account.

As mentioned, states are expected to invest all necessary resources to immediately realise the minimum core obligations of the human right to water. However, there might be states which are unable to immediately accomplish this. Nevertheless, in most cases the reallocation of resources can assist states in universally realising the minimum core obligations.[923] In order for this reallocation to take place, the distribution of water must be renegotiated.

Assessment criteria: Which trade-offs have to be made? How are the costs and benefits distributed?

7. Financial arrangements

When it comes to financial arrangements regarding a sustainable human right to water for vulnerable groups, two aspects deserve special attention. Firstly, financing and budgeting for the realisation of this sustainable human right from the perspective

921 Andreas Fischer-Lescano and Gunther Teubner, 'Regime-Collisions: The Vain Search for Legal Unity in the Fragmentation of Global Law' (2004) 25 Michigan Journal of International Law 999. p. 1039.
922 Stenekes. p. 15.
923 Winkler, *The Human Right to Water Significance, Legal Status and Implications for Water Allocation*. p. 122.

of the service provider or the state, which bears final responsibility. Secondly, what is extremely relevant for vulnerable groups in relation to financial arrangements is the affordability of access to water.

In developing their budgets, states are obliged to allocate resources for the realisation of human rights. They nevertheless have a margin of appreciation as they enjoy the right to self-determination.[924] As mentioned, even though the Covenant does not include a benchmark, meaning that there is no fixed percentage of the budget which must be allocated, so it is presumed that states are required to show that they have adequately considered the financial resources available in order to satisfy the human right to water.[925] The 2006 UNDP Human Development Report suggests that states should spend a minimum 1% of their Gross Domestic Product on water and sanitation.[926]

If a state has insufficient financial resources to realise this right, it must seek international cooperation and assistance. The UN advises that 0.70 percent of the Gross National Product of developed states should go towards international cooperation in order to meet the needs of populations in developing states.[927] They can for instance increase their resources by seeking donor financing or increasing revenue by progressive taxation.[928]

Besides government-raised financing (through, for instance, taxes) and transfers (such as grants from international organisation), contributions by households and users are other sources for funding water and sanitation services. These contributions are required to be affordable, including for vulnerable groups. Here it is important whether households are connected to informal or formal services. In the case of informal services, it is often challenging to guarantee affordability and to track developments.[929] Nevertheless, states still have final responsibility when it comes to ensuring the affordability of services.

In setting connection charges and tariffs, states must carefully consider certain aspects. Firstly, when the costs are too high, households which cannot afford them will rely on other sources or find ways to illegally gain access. Secondly, if the tariffs are too low, meaning below the affordability standard, the revenues might be too low and will require additional investment by states. Also, the low cost of services might also encourage wasteful water use which is contrary to the principle of sustainability. Another form of sustainability is that of financial sustainability, Thirdly, it is advisable

---

924 Marco Odello and Francesco Seatzu, 'The International Covenant on Economic, Social and Cultural Rights', *The UN Committee on economic, social and cultural rights: the law, process and practice* (Routledge 2013). p. 17 & Catarina De Albuquerque and Virginia Roaf, 'Financing', *Realising the human rights to water and sanitation: A handbook by the UN Special Rapporteur Catarina De Albuquerque* (2014). p. 7.

925 Philip Alston and Gerard Quinn, 'The Nature and Scope of States Parties' Obligations under the International Covenant on Economic, Social and Cultural Rights' (1987) 9 Human Rights Quarterly 156. p. 180.

926 United Nations Development Programme. p. 9.

927 (Odello & Seatzu, 2013) p. 19.

928 De Albuquerque and Roaf, 'Financing'. p. 7.

929 (De Albuquerque & Roaf, 2014b) p. 18.

that a regulatory body which is independent from the government is given the necessary authority to decide on tariffs as low tariffs are often used for political purposes in order to attract voters.

In order to make sure that relevant authorities set affordable tariffs it is advisable that states gather information on which percentages households spend on access to water. This information should be differentiated according to income and social groups. It also important to track where the household contributions are invested.

Assessment criteria: Is the financial arrangement for securing water for domestic purposes for vulnerable groups sustainable and equitable?

Implementation

Developing and implementing the sustainable realisation of the right to water are challenging tasks. There are several barriers to implementation such as the lack of political will, the lack of data or dependability on this data and reconsidering the allocation of over-allocated basins.[930]

8. Enforcement

Implementation, enforcement and monitoring are essential for the longevity of a sustainable human right to water. Not only are they necessary for upholding standards but also for evaluating whether the economic, social and ecological conditions have been met and whether adjustments and renegotiations are necessary.[931]

The enforcement of rights and policies is often a neglected aspect in water governance.[932] The emphasis lies mostly on the process of policy development, while implementation and enforcement are of equal importance. By articulating a sustainable human right to water in national legislation, the right moves from being a political and moral ambition to an enforceable right. Human rights are not self-enforcing and therefore enforcement mechanisms such as administrative and/or judicial enforcement have to be used for individuals and groups to assert their rights.[933] This is especially important for vulnerable groups as it can empower them directly or indirectly when they, or NGOs acting on their behalf, resort to the court system to enforce these rights.[934]

Even though formally laying down the human right to water in legislation is not necessary for the realisation of this right, the adoption of such a right can be highly beneficial for a number of reasons, as has already been explained.[935] However, adopting

---

930 Z Speed, R. Yuanyuan, L. Le Quesne, T. Pegram, G. Zhiwei, *Basin Water Allocation Planning: Principles, Procedures and Approaches for Basin Allocation Planning* (UNESCO 2013). p. 91.

931 See for instance Pienaar, Belcher and Grobler.

932 Van Rijswick and others. p. 12.

933 Ssenyonjo. p. 251.

934 Michael van der Valk and Penelope Keenan (eds), 'The Right to Water and Water Rights in a Changing World', *Colloquium the right to water and water rights in a changing world* (2011). p. 52.

935 See chapter 3.

this right is merely the first step in its realisation. What should be kept in mind is that human rights require additional measures for them to become enforceable. As mentioned, the enforceability of the human right to water has been the centre of the debate. Nevertheless, what can be said is that in order to increase its enforceability, regulations and policies need to specify aspects of the right such as standards for the quality of water. The elements of the sustainable human right to water should be translated into measurable indicators. This may prove a daunting task, but many governments and organisations have been instrumental in paving the way towards further defining the right and also making it context-specific.[936] However, the enforcement of a legal decision can still be challenging and depends on factors such as the capacity available to deal with this, the compliance mechanisms, such as fines, and the overall trust in the legal system.

Outside of judicial or administrative enforcement, monitoring is an additional mechanism for realising a sustainable human right to water. This will stimulate overall compliance and will assist those who do not assert their rights through enforcement mechanisms. States are required to monitor their compliance and that of others with the legal contents of the human right to water and other environmental obligations.[937] In order to facilitate this, indicators should be developed. The human rights framework requires structural, process and outcome indicators to be defined for monitoring.[938] States should therefore identify standards, targets and indicators regarding availability, accessibility, quality, affordability, acceptability, sustainability and non-discrimination.

Monitoring in itself is not sufficient as it is a tool for the further development of policy and also a tool for public awareness. Therefore, this building block is linked to, among other blocks, that of stakeholder participation. States must ensure 'that data are adequately collected, organised and stored, and then made public in a timely, accurate, accessible and useful form.'[939] Especially data concerning the inequalities in water access are important for realising the right for vulnerable groups. Often states do not supply disaggregated data which is useful for identifying where and how discrimination occurs and also whether progress is being made.

Assessment criteria: Are regulations and agreements enforceable by public and/or private parties, and are there appropriate remedies available?

9. Conflict prevention and resolution

Conflict prevention

Water is a finite resource and because of this the allocation thereof can be a source of conflict. In Water Law, the focus of conflict prevention is often on preventing

936 See for instance Winkler, 'Judicial Enforcement of the Human Right to Water – Case Law from South Africa, Argentina and India'.

937 De Albuquerque and Roaf, 'Monitoring'. p. 5.

938 Ibid. p. 7.

939 Ibid. p. 8.

conflicts between states and promoting cooperation among them in relation to shared watercourses by using mechanisms such as prior informed consent about planned measures.[940] However, also within states conflicts concerning water may occur. Conflicts in this sense mean a clash of opposing needs and interests and can arise at different levels and between different groups. For instance, there can be a conflict between different water uses but also physically between different groups.[941]

In relation to conflict prevention, particularly for vulnerable groups, sustainability and equity are highly relevant. Especially the principles of intergenerational and intragenerational equity are important. This is because with the exploitation of natural resources in many resource-rich developing countries, governments often gain substantial financial benefits. However, the social conditions of vulnerable groups are often neglected, threatening human security, which can lead to (armed) conflict between governments and local populations.[942] Because of this the UNDP advises that in order to prevent conflicts and to strive for human security, a balance must be found between the environment and the equitable distribution of resources while still considering the needs of current and future generations.[943]

Overall, the assumption is that having clear rules based on shared responsibilities and values, which are represented in several other building blocks, in alignment with the physical circumstances can in many cases prevent conflicts from arising. These values and responsibilities should however be grounded on equity and sustainability.

Conflict resolution

In the event that conflicts do occur, the presence of mechanisms which can deal with these conflicts in an independent manner are necessary. Frequently, independent mediators, arbiters or judges are chosen for this purpose. The absence of these mechanisms adversely impacts and excludes affected vulnerable communities.[944] As mentioned in the building block on participation, access to justice is a necessary element in conflict resolution, especially for vulnerable groups. Tied to this are the principles of public participation and access to information. Many governments have developed laws and regulations regarding access to justice. However, the implementation of these laws is often insufficient.[945] For this reason it is not only relevant to assess whether the appropriate laws and regulations are in place but also whether the outcomes of these mechanisms are being enforced and implemented.

---

940 Joyeeta Gupta and Joseph W Dellapenna, 'The Challenges for the Twenty-First Century : A Critical Approach' in Joseph W; Dellapenna and Joyeeta; Gupta (eds), *The Evolution of the Law and Politics of Water* (2009th edn, Springer Science 2009). p. 399.

941 Misiedjan, van Rijswick and Tjen A Kwoei.

942 Marong. p. 38.

943 UNDP developed the theory of sustainable human development see Human Development Report.

944 (Krchnak, 2004) p. 34.

945 Karin M Krchnak, 'Improving Water Governance Through Increased Public Access to Information and Participation' (2004) 5 Sustainable Development Law & Policy 34. p. 34.

What should also be kept in mind is that judicial mechanisms often apply to individual cases only. However, if the issue is a problem which is relevant for a bigger demographic than just the individual parties to a case, mechanisms should be in place for the issue to be taken up by other authorities. For instance, if the water allocation mechanism in place has become inappropriate due to a significant change in water quantity and a dispute occurs because of this, other relevant authorities should pick up on this issue leading possibly to necessary policy and legal changes.

Assessment criteria: Are there sufficient conflict prevention and resolution mechanisms in place?

10. Engineering and monitoring

When it comes to technology and engineering the choices need to be appropriate and economically and socially viable. Sustainable services and the technology involved therein may require higher investment costs and relatively more maintenance. For this reason choices concerning technology must be made based on sufficient information explaining existing and projected resources in the medium and long term.

One of the dilemmas that states face when resources, especially water resources, are scarce is the prioritisation of sustainability or extending services to those who are currently unserved.[946] What states should consider is that from the human rights perspective states must first strive for equality through the efficient use of resources. States should expand access to minimum essential services before improving the services for those who are already served. In this sense the principles of equality and sustainability complement each other.[947]

Sustainable infrastructure

Substantive technology choices can specify the tools which can be applied: 'methods to prioritise and analyse the financial viability of investments, the role of subsidies, efficiency incentives, technologies/methods to analyse and account for resources and reserves as well as water uses, to increase water use efficiency, to obtain non-conventional resources, to control water use and analyse ecological flows as a basis for improved decision-making'.[948]

> *'Realizing the right to water is often constrained by the limited coverage of network systems. There is a need to implement solutions at the community and household levels which can*

---

946 De Albuquerque and Roaf, 'Principles'. p. 78.

947 Ibid. p. 78.

948 'Governments Contribution to the Implementation of the Sustainable Development Goals Related to Water | 2015 UN-Water Annual International Zaragoza Conference. Water and Sustainable Development: From Vision to Action. 15–17 January 2015' <www.un.org/waterforlifedecade/waterandsustainabledevelopment2015/stakeholders_governments_and_local_authorities.shtml> accessed 5 February 2016.

provide safe drinking water using appropriate and cost-effective technologies. However, the implementation of these technologies needs to be supported by awareness raising and capacity building at the household and community levels.'[949]

Assessment criteria: Are the technology choices appropriate?

6.4 CONCLUSION

Developing an integrated assessment method for vulnerable groups' sustainable access to water for domestic purposes is a challenging exercise. This method highlights the main topics which assist in reviewing national water management and the realisation of the right to water for vulnerable groups in a systematic manner. The method incorporates information from different disciplines to paint a clearer picture of the overall situation and not merely looking at the regulatory system. Access to water for domestic purposes is not merely influenced by whether or not the human right to water is recognised but other circumstances also influence this access.

The building blocks are interlinked and point out the three different aspects: content, organisation and implementation. This emphasises that not only the outcome is important but also the process, especially for vulnerable groups whose interests are often not taken into consideration. This assessment framework will be used in the following chapter which reviews access to water for vulnerable people in Suriname.

---

949 Www.un.org/waterforlifedecade/waterandsustainabledevelopment2015/pdf/OP_Governm_WASH_Pireh_FORMAT.pdf.

CHAPTER 7
SURINAME

"We, the people of Suriname, value and protect our biological diversity, including all natural and cultural resources, through equitable and sustainable use for present and future generations. A national commitment to Suriname's biological wealth, integrating all sectors of society, will allow people to experience the full benefits of sustainable management and wise use while maintaining and enhancing the diversity of the country's cultural and natural heritage."[950]

7.1 INTRODUCTION

The previous chapters laid out which obligations states have towards vulnerable individuals and groups in relation to their to access water. The protection of water resources plays an important role in this but also the (re)allocation of these resources is an important factor as well. The standards have been set at the international level through soft and hard law instruments.[951] However, for individuals and groups it is most relevant how and whether states have implemented these obligations at the national level. Therefore, this study includes case-studies to explore whether and how international obligations and other standards influence how a sustainable human right to water is shaped for vulnerable groups and individuals and how water resources should be managed. To demonstrate this the Republic of Suriname was selected. The two cases (Paramaribo and Brokopondo) which are set in this state show how access to water is managed and what the different challenges are. They explore the relationship between the human right to water and how water resources are managed. Figure 7.1 which was introduced in chapter 5 illustrate this relationship. The protection and management of water resources form the outer boundaries of the realization of the human right to water.

950 The Ministry of Labour Technological Development and Environment, 'The Fourth National Report to the Convention on Biological Diversity' (2013). p. 9.

951 See chapter 4.

Figure 7.1 Representation of the distant corollaries and core of the human right to water

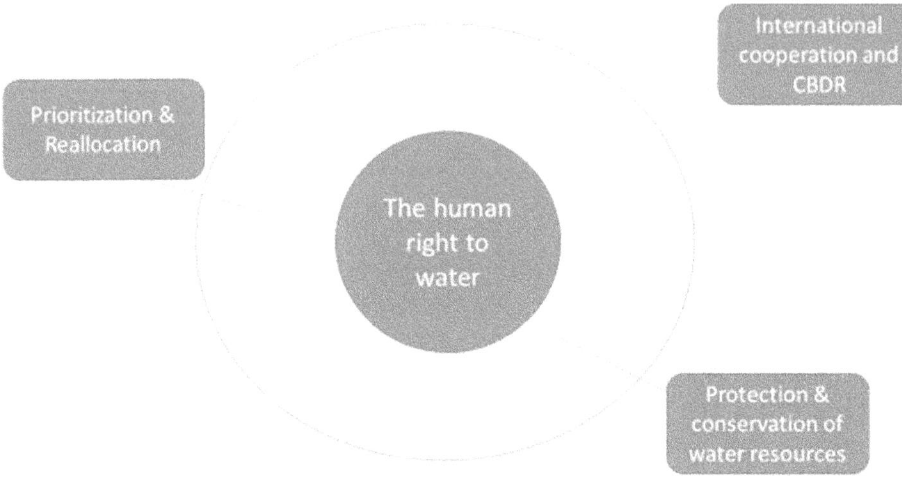

Suriname has a great wealth of water resources, so much so that it is in the top 10 of water-rich countries in the world.[952] Therefore it can be assumed that Suriname has the necessary natural resources to provide all of its inhabitants, including vulnerable groups, with access to safe water. However, not every inhabitant of Suriname is currently guaranteed access to safe drinking water, especially those living outside of the urban areas, due to regional discrepancies in coverage.[953] According to the Ministry of Labour, Technological Development and Environment threats to the freshwater ecosystems are: '(1) pollution (urban-domestic and industrial waste) e.g. mining activities, (2) changes in land use, (3) agriculture runoff (e.g. pesticides), (4) climate change (droughts, floods), and (5) sea level rise (saltwater intrusion).'[954] From a human rights perspective states should follow the authoritative interpretation of the human right to water to ensure that everyone has 'sufficient, safe, acceptable, physically accessible and affordable water for personal and domestic purposes.'[955] Environmental requirements can support these efforts for the realisation of a sustainable human right to water. Suriname is visibly experiencing challenges in bringing this about and this research aims to increase understanding about these challenges and to contribute to possible solutions. Therefore the research question of this chapter is the following: How can a sustainable human right to water for vulnerable groups be fulfilled in Suriname?

---

952 Alper Baba and others (eds), *Climate Change and Its Effects on Water Resources* (Springer Science Business Media BV 2011). p. 286.

953 See section 7.3.

954 The Ministry of Labour Technological Development and Environment. p. 22.

955 Committee on Economic Social and Cultural Rights, 'General Comment No. 15 The Right to Water' (2002). paragraph 2.

In order to find solutions this chapter explores the major obstacles to the enjoyment of a sustainable human right by vulnerable groups in Suriname by discussing two case studies which tie into the aspects which have been identified in Chapter 5: prioritisation and reallocation, protection and conservation of water resources. This analysis will be done systematically by applying the '10 building block' method explained in the previous chapter. The chapter begins by explaining the context of Suriname. In section 7.3 the methodology and its limitations are discussed. This is followed by the introduction to the case-studies and an overview of the state of the human right to water in Suriname. Afterwards the building-block method is applied to the case studies. The chapter will conclude with recommendations for improving the situation of the different building blocks.

7.2 BACKGROUND TO SURINAME

The Republic of Suriname lies on the north-eastern coast of South America and shares its borders with Guyana to the west, French Guyana to the east and Brazil to the south. The Atlantic Ocean forms the northern border. The other borders shared with neighbouring states are natural borders shaped by the water bodies, the Corantijn River, the Marowijne River and the Catchment divide of the Amazon Basin. Suriname is the smallest mainland country in South America and still has the most untouched Amazon rainforest.[956] Suriname has a tropical climate where four seasons can be distinguished: 'the short rainy season from early December to early February; the short dry season from early February to late April; the long rainy season from late April; to mid-August; and the long dry season from mid- August to early December.'[957]

The population is approximately 541,638 with most of the population living along the coast.[958] The total land area is 163,830 km^2 which makes the population density approximately 3.1 inhabitants per km2.[959] There are ten administrative districts: Brokopondo, Commewijne, Coronie, Marowijne, Nickerie, Para, Paramaribo, Saramacca, Sipaliwini and Wanica. The population density is the highest in Paramaribo, Wanica and Commewijne. What should be kept in mind, however, is that during the last decade large groups of immigrants have come into Suriname mostly from countries in the region like Guyana, French Guyana, Brazil and Venezuela and from outside of the region, mainly China.[960] A significant number of these immigrants have remained

956 The Amazon rainforest covers 80% of the Surinamese territory Marieke Heemskerk, 'Livelihood Decision Making and Environmental Degradation: Small-Scale Gold Mining in the Suriname Amazon' (2002) 15 Society & Natural Resources 327. p. 329.

957 N.V. Surinaamse Waterleiding Maatschappij, 'Suriname Water Supply Master Plan'. p. 3.

958 Algemeen Bureau voor de Statistiek – Suriname, Conservation International and UNDP, 'Environment Statistics' (2014). p. 8.

959 The Ministry of Labour Technological Development and Environment. p. 10.

960 It is reported that at least 10,000 Chinese people migrated to Suriname between 2000 and the end of 2007.
see: http://nos.nl/artikel/215759-china-houdt-van-suriname-maar-t-is-niet-wederzijds.html.

unregistered. It is for instance estimated that between 20,000 and 40,000 Brazilians have settled in Suriname.[961] This number is however not visible in the official statistics. The human rights of these groups also need to be respected and realised by Suriname. This obligation lies on Suriname.[962]

Table 7.1 Demographic per district

District[963]	Total	%
Paramaribo	240,924	44,5
Wanica	118,222	21,8
Urban	**359,146**	**66,3**
Nickerie	34,233	6,3
Coronie	2,887	0,6
Saramacca	17,480	3,2
Commewijne	31,420	5,8
Para	24,700	4,6
Rural	**111,224**	**20,5**
Marowijne	18,294	3,4
Brokopondo	15,909	2,9
Sipaliwini	37,065	6,8
Interior	**71,268**	**13,1**
Total	**541,638**	**100**

Suriname gained independence from the Netherlands after being under European colonial rule since the 17th century. However, ties between the Republic of Suriname and the Netherlands remain to this day, economically, legally, culturally and linguistically, to name but few. However, the strength of these ties is dependent on political sentiments.[964]

Even though Suriname is the smallest country on the mainland in South America, it houses a multitude of ethnic, religious and linguistic groups due to its colonial past. Suriname was inhabited by numerous indigenous tribes before it was conquered by colonial powers in the 17th century. These colonial powers transported other groups such as enslaved Africans, indentured workers from India and Indonesia, prosecuted Jews from Spain and Portugal and colonial settlers from England and the Netherlands to Suriname over a long period of time. As a result, Suriname now has a population

961 Marjo de Theije, ''De Brazilianen Stelen Al Ons Goud!' Braziliaanse Migranten in Stad En Binnenland' (2007) 26 OSO, Tijdschrift voor Surinamistiek en het Caribische Gebied 81. p. 81.

962 As explained in chapter 3.

963 Algemeen Bureau voor de Statistiek – Suriname, Conservation International and UNDP. p. 8.

964 Roger Janssen, *In Search of a Path An Analysis of the Foreigh Policy of Suriname from 1975 to 1991* (KITLV Press 2011).

where the largest ethnic group is of Indian descent (29%), followed by Creoles (24.1%) and the Javanese (14.6%). Other groups are the Maroons[965] (14.7%) and the Indigenous Population (3.6%).[966] This diversity led to over 61 languages being spoken in Suriname.[967] Suriname has a complex relationship between culture, ethnicity, class, and gender.[968]

Approximately 70 to 85 percent of the Surinamese population live below the poverty line. Those living in the Interior are overrepresented in this category.[969] Yet, the country is characterised as an upper-middle income country by the World Bank.[970] Unemployment is a major challenge, especially amongst young people. And the Government is the biggest employer in Suriname, proving employment to 60 percent of the working population.[971]

The Surinamese economy very much depends on natural resources, mostly gold and timber.[972] The economy experienced growth between 2004 and 2014 with an average of 4.5 percent per year. This growth ceased in 2015 as the currency greatly depreciated accompanied by high inflation rates.[973] As a result the Surinamese government initially accepted assistance for its economic reform program from the World Bank and other international financial institutions. Part of this reform program is the reduction of subsidies, the lowering of public spending, private sector development and the introduction of value-added tax.[974] So far, Suriname is not on the road back to economic stability or growth. As of March 2017, Suriname is experiencing high inflation.[975] And in May 2017, Suriname revoked its agreement with the IMF.[976]

Suriname's legal system can be qualified as weak if measured by the rule of law. This is the result of 'the slowness and delays in the judicial process, combined with a large informal sector that largely evades paying taxes and general concerns about favouritism and corruption in public sector decision-making processes.'[977] In addition, Suriname has a limited public and administrative regulatory framework. As a result, administrative instruments such as administrative enforcement are lacking in many areas and also access to a constitutional court, even though this is laid down in the

---

965 Descendants of self-emancipated enslaved Africans who established independent communities in the Interior during.

966 'Milieustatistieken Environment Statistics'.

967 The Ministry of Labour Technological Development and Environment. p. 10.

968 Rosemarijn Hoefte, *Suriname in the Long Twentieth Century: Domination, Contestation, Globalization* (2013) <http://books.google.com/books?id=sHdanQEACAAJ&pgis=1>. p. 11.

969 ABS 2009 also see http://dwtonline.com/laatste-nieuws/2016/07/21/armoedegrens-wordt-eindelijk-vastgesteld/

970 Www.worldbank.org/en/country/suriname/overview.

971 ABS 2005.

972 In the past, bauxite was a great source of income. However, since 2010 the bauxite production has stalled.

973 Www.worldbank.org/en/country/suriname/overview.

974 Www.iadb.org/en/countries/suriname/suriname-and-the-idb,1032.html.

975 Www.starnieuws.com/index.php/welcome/index/nieuwsitem/41062.

976 Http://nos.nl/artikel/2173577-suriname-ziet-af-van-rest-imf-lening.html.

977 Executive summary International Development Bank, 'Governance in Suriname' (2001).

Surinamese Constitution, is non-existent.[978] Lower courts are allowed to review concrete cases against political and civil rights. However, they may not declare any regulation to be generally non-binding.[979]

The sections above provide a short overview of the situation in which the human right to water is supposed to function. As mentioned before, how the human right to water is implemented within a national context very much depends on the circumstances. Water management is dependent on the physical circumstances but it is also shaped by historical developments and population dynamics. The following sections will discuss the current status of the human right to water in Suriname and the context in which this right has to be implemented.

7.2.1 THE SITUATION OF THE HUMAN RIGHT TO WATER IN SURINAME

To paint a clear picture of the situation of the human right to water in Suriname and to assess for how many people the right has been realised is challenging. This is because most of the available data explains how many people in Suriname have access to water for domestic purposes. This data is however often inconsistent due to the use of different methods of analysis, different measurements and different sources. As a result, the studies mention different percentages and often the measurements are not regularly followed up on, making the data unreliable. This is mostly caused by research being grounded on different premises. For example, a study conducted by UNICEF in 2010 states that 70,7 percent of the Interior has access to safe drinking water. However, this study is a 'representative sample survey of women aged 15–49 and children under age five of 7,407 responding households out of a total of 9,356 sampled households. The main purpose of the Multiple Indicator Cluster Survey (MICS) 2010 is to support the government of Suriname to generate statistically sound and comparable data for monitoring the situation of children and women in the country.'[980] According to the 2014 census, 71,268 individuals were recorded in the Interior. As households comprise of around 4 individuals on average it could be concluded that the 2010 study by UNICEF only covers around 42 percent of the total number of households. For this reason, this data is not completely representative of the overall situation in the Interior of Suriname, but it does give an indication.

Other documents which provide indications are the national MDG Reports. In the 2009 MDG Progress Report, the target of access to safe drinking water was estimated to having the potential of being reached while the supportive environment was said to

978 Article 144(2) Constitution.

979 Nancy del Prado, 'Analysis of Instruments, Principles and Legal Norms in Force on Domestic Law of the Amazon Basin in Water Resources Management' (2013). p. 15.

980 Abstract http://microdata.worldbank.org/index.php/catalog/1913/study-description.

be weak but improving.[981] However, the 2014 MDG Progress Report merely mentioned that 97.3 percent of the population have sustainable access to safe drinking water.[982]

This study refers to the data from three studies. Firstly, the 2005 Suriname MDG Baseline Report because it does differentiate between districts and this book focusses on access to water in two different districts.[983]

Table 7.2 Percentage of population with access to safe drinking water sources[984]

District	%
Urban	92.6
Rural	66.6
Interior	20.0
National average	72.6

Secondly, the data from the household census conducted in 2014 by the General Bureau of Statistics.[985]

Table 7.3 Overview household main water supply[986]

	Total	% of total	Male	Female
Water connection in house	83,252	59.31	55,708	27,544
Water connection outside the home < 200 meters	13,226	9.4	8,520	4,706
Water connection outside the home > 200 meters	2,500	1.78	1,534	966
Water collection in receptacle	26,876	19.15	18,519	8,357
Well	1,843	1.32	1,433	410
Creek or river	3,222	2.29	1,280	1,942
Water truck	385	0.03	253	132
Bottled water	3,416	2.43	2,552	864
Other	1,980	1.4	1,407	573
Unknown	18,294	13	2,315	1,352
Total	**140,367**	**100**	**93,521**	**46,846**

981 Government of Suriname, 'MDG Progress Report 2009' (2009). p. 74.

982 Government of the Republic of Suriname, 'MGD Progress Report 2014' (2014). p. 100.

983 Ibid. p. 41.

984 Government of Suriname and United Nations Country Team Suriname, 'Suriname MDG Baseline Report' (2005). p. 41.

985 Algemeen Bureau voor de Statistiek – Suriname, 'Resultaten Achtste Volks- En Woningtelling in Suriname (Volume III) Huishoudens, Woonverblijven En Gezinnen, Milieu, Criminaliteit' (2014). p. 69; 2012 National census, of 140,367 households demographic, economic and social data were recorded.

986 Ibid. p. 69; 2012 National census, from of 140,367 households demographic, economic and social data was recorded.

And, thirdly, the access rates made available by the Suriname Water Maatschappij (SWM) in its Master Plan.[987] This shows how many households are connected to the general water supply system. Those not connected make use of the services of DWV or take measures into their own hands.

Table 7.4 Access of dwellings to piped water supply per coastal area district (%)[988]

Access per District	No. (%) of dwellings connected
Wanica	60
Paramaribo	92
Nickerie	91
Saramacca	62
Commewijne	35
Marowijne	49
Para	62
Coronie	94
Brokopondo	11

As explained in chapter 3, the human right to water does not only require physical accessibility. It also requires that other conditions are met such as economic accessibility and cultural appropriateness and standards for the quality and quantity of water. The data available on the situation in Suriname only provides an insight into how many individuals or households are connected to the water supply infrastructure or receive water services through other means. The data does not provide information on whether these households and individuals receive water on a continuous basis or on the quality of the water that they receive. Also the data is not disaggregated by sections of society and therefore does not show which sections of society experience specific challenges in securing their access.

7.3 METHODOLOGY AND LIMITATIONS

As mentioned, the classic legal method entails analysing the conventional sources of law through desk research. An analysis of policy documents and academic literature has also been applied for finding relevant information for the building blocks. However, this method by itself was not sufficient for understanding the major challenges to the

987 N.V. Surinaamse Waterleiding Maatschappij, 'Suriname Water Supply Master Plan'. p. 19.

988 N.V. Surinaamse Waterleiding Maatschappij, 'Suriname Water Supply Master Plan'. p. 19. Note #1: The access rate is only for the urbanized areas, while some of the districts also cover Hinterland areas. In Brokopondo 50% of dwellings have access to piped water supply with distances over 200 m. Note #2: Sipaliwini District is considered to be Hinterland. Note #3: Figures based on the ABS 2004 census data.

enjoyment of a sustainable human right to water by vulnerable groups in Suriname. Therefore, social science methods were included on a limited basis. Stakeholder and expert interviews were conducted and other data and documentation from other disciplines were studied. Information was collected through a number of interviews with government officials, political leaders, NGOs, communities, and other relevant parties. Combining these two methods was necessary as relevant data on the situation in Suriname is not readily available or is often inconsistent. One of the reasons for this is a lack of digitalisation which necessitates access to hard copy material and this gave rise to several trips to Suriname in order to collect the necessary information.

Table 7.5 Dimensions for the assessment of sustainable access to water for vulnerable groups

Preliminary research was conducted in Suriname during a period of three months from March to June 2015. A second visit to Suriname took place from May to June 2016. These trips included field trips to Nickerie, Paramaribo and two villages in the Brokopondo district, Asigron and Balingsoela. To receive the necessary feedback on preliminary ideas, several events where organised where stakeholders such as government agencies and individuals were able to express their ideas and to convey their knowledge. And finally, in 2017 there was a two-week trip to Suriname.[989] In the latter context, a field trip to Asigron was made together with government agents. A course on water governance was given to participants from several ministries, governmental agencies and the private sector.[990]

989 Financial support for this trip was provided by the UNDP.

990 See annex A for a list of participants of the different events and an overview of activities can be found. The first two trips were financed by the Silvia W. de Groot Fonds. The last visit to Suriname was partly financed by the UNDP, the WWF, the Ministry of Natural Resources and Waterforum Suriname.

7.4 INTRODUCTION TO THE CASE STUDIES

The inclusion of case studies allows for the human right to water to be localised. What the case studies show is that the realisation of the human right to water requires different measures in different situations. This is because the interferences with the right to water differ. The following sections provide a further introduction into the two case studies, two specific districts within Suriname.

7.4.1 PARAMARIBO: REALISING SUSTAINABLE ACCESS TO WATER FOR DOMESTIC PURPOSES FOR POOR AND UNSERVED COMMUNITIES

Suriname's capital, Paramaribo, has the highest national coverage regarding access to water with around 92 percent of the city's inhabitants have access.[991] However, this still means that 1 in 10 persons in Paramaribo is not connected to the public water supply system. Great disparities also exist between neighbourhoods.[992] Often, those who are unserved live in informal housing. Some of these communities are located on the outskirts of Paramaribo. The residents either live in informal settlements or are illegally occupying property meant for social housing.[993] These informal settlements were established after the Interior War[994] which took place between 1986 and 1992 and claimed the lives of hundreds of people while thousands fled from the Interior into the capital or other places.[995] The neighbourhoods Sunny Point and Texas are examples of such communities. The communities continued to live there and consecutive governments have not provided alternative housing.[996] They are often unable to depend on the government for public services. These communities mostly consist of Maroon communities who used to live in self-sustaining communities in the Interior. For instance, after the Interior War a community originating from the Sarammaccan village of Pokigron, settled in a refugee camp behind the well-known sports stadium, Anthony Nesty Sporthal, and lived there in informal housing until approximately 2015 when they were granted plots of land by the government.[997] As explained, when

991 Government of Suriname and United Nations Country Team Suriname. p. 41 & N.V. Surinaamse Waterleiding Maatschappij, 'Suriname Water Supply Master Plan'. p. 19.

992 'Among resorts (neighbourhoods) within Paramaribo and Wanica, access to tap water ranges between 25.2% and 97.7% of households. Among these same resorts, the percentage of the population with a tertiary education ranges from 1.5% to 10.7% (5). Poverty-stricken urban areas have indicators very similar to the rural interior.'
PAHO/WHO, 'PAHO/WHO Country Cooperation Strategy Suriname 2012 – 2016' (2012). p. 11.

993 Gloria Sandie, 'Sunny Point Access to Water as a Human Right in the Coastal Region of Suriname' (FHR Lim A Po Institute 2014).

994 Ellen de Vries, *Suriname Na de Binnenlandse Oorlog* (LM Publishers 2005). p. 86.

995 Hoefte. p. 24.

996 De Vries p. 81.

997 Remak Jeroe (39), one of the one hundred or so inhabitants of Achter Nis, knows all about this. For 15 years the former inhabitants of the Marron village Pokigron, which was destroyed during the war

communities are not connected to the public water supply they depend on alternative sources for their water, such as rain water harvesting.[998]

Next to those who are currently unserved, lower income households which are connected to the public water supply are also at risk of not having their human right to water sustainably realised. In the context of Suriname, the risks are caused by issues such as decreasing water quality, low water pressure and increasing water tariffs in conjunction with the overall economic crisis. In addition, the Surinamese government decided in August 2015 to cut subsidies for not only the public water supply but also electricity and gas.[999] The quality and quantity of freshwater used for the public water supply have been compromised due to the intrusion of saltwater in groundwater reservoirs. Appropriate alternative sources of freshwater are therefore being sought. Another issue is that the water supply infrastructure is in need of modernisation and replacement. The current condition of the system is susceptible to leakages and can cause health problems. This case-study focusses on unserved and underserved communities living in informal settlements or have occupied housing as well as lower income households. The overall condition of the human right to water in Paramaribo will also be discussed as it affects the general population.

7.4.2 BROKOPONDO: MINING AS A MAJOR THREAT TO ACCESS TO WATER FOR DOMESTIC PURPOSES IN THE INTERIOR

Brokopondo is a district located towards the centre of Suriname and is part of what is known as the Interior of Suriname.[1000] Brokopondo has around 16,000 inhabitants[1001] and is divided into six resorts, Brownsberg, Centrum, Klaaskreek, Kwakoegron, Marchalikreek and Sarakreek. An overwhelming majority of the population in Brokopondo belong to indigenous and tribal communities[1002], with the largest group being the Saramaka people. Brokopondo is of great significant to the (economic) development of Suriname for several reasons. Firstly, gold was found here in the 21st century and is currently explored by private companies and small goldminers. Suriname is highly dependent on the revenues from gold and oil.[1003] Secondly, the Brokopondo reservoir (officially known as Prof. Dr. Ir. W.J. Van Blommensteinmeer) is

between Ronnie Brunswijk and the army, have lived in pitiful conditions in a camp just behind the modern Anthony Nesty Stadium in Paramaribo. There is no electricity and no water. The families live in gloomy and dark rooms.
www.volkskrant.nl/archief/stemmen-in-suriname~a644721/

998 Jurna p. 51.

999 Http://nos.nl/artikel/2055390-suriname-verdrievoudigt-prijs-van-elektriciteit.html.

1000 With the Interior Brokopondo is not always classified as being part of the 'Interior.' For this study the Interior consists of the districts: Marowijne, Brokopondo and Sipaliwini.

1001 Algemeen Bureau voor de Statistiek – Suriname, Conservation International and UNDP p. 8.

1002 Ellen-Rose Kambel and Fergus MacKay, *De Rechten van Inheemse Volken En Marrons in Suriname* (KITLV Uitgeverij 2003). p. 105.

1003 Https://data.worldbank.org/country/suriname.

located in this district which provides several large industries and the coastal area with electricity.[1004] Mining predominantly takes place in the Interior of Suriname where only 20 percent of the population has access to safe water for domestic purposes and therefore 80 percent depend on traditional sources of water such as, rivers, creeks and rain-water collection, to foresee in their needs. Unfortunately, mercury used for gold mining, other by-products of industries and (human) waste end up untreated in the surface water and contaminate water ways, damaging ecosystems and making surface water unsuitable for domestic purposes.

The negative impacts of especially mercury pollution and limited access to water safe for domestic use are also being felt in other regions. For instance, the Marowijne River, along which several indigenous and tribal communities live, is polluted with mercury. Up to 79% of the children from these communities have abnormally high mercury levels in their hair which negatively affects their (neurological) development.[1005] They also run the risk of experiencing itchiness, sore eyes and rashes after bathing in the water.[1006] In addition to mercury poisoning, according to the Medical Mission, the most prevailing health problems in the Interior are problems regarding hygiene, caused by restricted access to sanitary facilities as there are no latrines available in most villages and the communities collect their water from creeks and rivers.[1007]

As explained, Suriname has the legal responsibility to facilitate the enjoyment of rights and to protect the environment, especially when vulnerable groups depend on this. The gold mining sector has been detrimental to the environment in Suriname. This sector can be divided into two categories: large scale mining and artisanal or small scale mining. In general mining has harmful and far-reaching environmental effects such as deforestation, and the loss of wildlife and fauna.[1008] And here lies the dilemma. Small-scale mining is a form of subsistence mining which provides a direct livelihood for the miners who often belong to tribal communities. In Suriname, this form of mining often takes place alongside large-scale operations. Mining and logging concessions overlap with the territories and villages of tribal and indigenous communities.[1009] The families of the miners live in these communities and are most at risk of experiencing the negative effects of small-scale goldmining.[1010]

1004 Electricity to the Interior is provided in the form of diesel generators. See Algemeen Bureau voor de Statistiek – Suriname, Conservation International and UNDP. p. 98.

1005 D Wip and others, 'Urban Mercury Pollution in the City of Paramaribo, Suriname' (2013) 6 Air Quality, Atmosphere and Health 205. p. 206.

1006 James Harnum, 'Deriving the Right to Water from the Right to Life, Liberty and Security of the Person : Section 7 of the Canadian Charter of Rights and Freedoms and Aboriginal Communities in Canada' (2010) 19 Review of European Community and International Environmental Law 306. p. 308.

1007 Government of Suriname and United Nations Country Team Suriname. p. 31.

1008 Catarina De Albuquerque and Virginia Roaf, 'Monitoring', *Realising the human rights to water and sanitation: A handbook by the UN Special Rapporteur Catarina De Albuquerque* (2014). p. 34.

1009 (Heemskerk & Kooye, 2003) p. 5.

1010 (Heemskerk & Kooye, 2003) p. 14.

Between 20,000 – 25,000 small scale goldminers[1011] or, according to some estimates, even twice as many are active in Suriname in areas such as Brownsberg (Brokopondo) and East Suriname.[1012] Maroon communities have traditionally mined for gold as it has been part of their culture even before the transatlantic slave trade.[1013] A large part of the gold extracted in Suriname is extracted by small-scale gold miners. Surinamese maroons make up a quarter of the mining population and Brazilians (so-called: Garimpeiros) make up the rest.[1014] The Brazilian gold miners which are often undocumented and other illegal miners have settled in different areas of the Interior and have formed communities. This research focusses on tribal and indigenous communities in the Interior.

7.4.3 JUSTIFICATION FOR CASE STUDIES

As the two case studies demonstrate, Suriname is confronted with both pollution as well as depletion and this requires different strategies which can be region-specific. The two case studies reflect this and concern the issue of water services (Paramaribo) and the protection of the water system (Brokopondo). Because the second case study discusses the direct dependence on the water system as universal infrastructural coverage is lacking in the Interior, this case study also demonstrates the importance of taking into account the environmental vulnerability of individuals and groups. The next part of this chapter systematically analyses the building blocks for the case studies. They will be discussed one after the other.

7.5 APPLICATION OF THE TEN BUILDING BLOCKS FRAMEWORK

In the following sections the relevant aspects of the case studies will be discussed according to the building blocks. For each building block several questions have been selected which assist in analysing the strength of the specific building block. Each building block discussion starts off with general comments which are followed by an analysis of relevant aspects for the blocks relating to respectively Paramaribo and Brokopondo. Some of the building blocks may have more relevance than others because in some situations

1011 Small-scale gold mining is characterized its informality and a not formally trained labour force. (Heemskerk & Kooye, 2003) p. 1.

1012 Marieke Heemskerk and Rachael Van Der Kooye, 'Challenges To Sustainable Small-Scale Mine Development in Suriname' in Gavin Hilson (ed), *The Socio-Economic Impacts of Artisanal and Small-Scale Mining in Developing Countries* (Taylor and Francis 2003). p. 1.

1013 Interview with Christopher Healy (Culterecom)
See EA Ofosu-Mensah, 'Historical Overview of Traditional and Modern Gold Mining in Ghana' (2011) 1 International Research Journal of Library, Information and Archival Studies 6. for an historical overview of tradition and modern gold mining in Ghana.

1014 Heemskerk and Kooye. p. 6.

information might be lacking or limited. Thus not each building block will go into specific local aspects. After the discussion of the ten building blocks, a concluding overview will be provided illustrating the strengths and weaknesses of the Surinamese system which supports the realisation of the human right to water for vulnerable groups.

7.5.1 BUILDING BLOCK 1: WATER SYSTEM KNOWLEDGE

Assessment criteria: Is there sufficient knowledge of the existing water system in order to support access to water for vulnerable groups? If not, what are the gaps; is sufficient knowledge available to assess the impact on the water system because of changes in the environment and societal functions?[1015]

National level

Knowledge of the water system in Suriname concerns both knowledge about the natural physical resources as well as the man-made infrastructure, such as dams. Suriname has a great diversity of water bodies. The interior is mountainous with a dense network of streams. Suriname's main source of freshwater is its annual rainfall of 2200 mm[1016] resulting in many streams and large wetlands. There are seven main rivers which originate in the inlands of Suriname. These rivers carry 4,800 m3 of freshwater to the Atlantic Ocean. This makes up about 30% of the annual rainfall.[1017] The remaining rainfall either evaporates or replenishes groundwater reserves.[1018] These groundwater reserves can be found along the coast. The use of freshwater in Suriname is estimated to be similar to that of other developing and middle-income countries, meaning: agricultural use accounts for 82 percent of water use while the industrial sector uses 10 percent and domestic use is around 8 percent.[1019]

Paramaribo

In order to provide Paramaribo and other coastal cities with water for domestic purposes ground water is used. It has been determined by the SWM and technical

1015 Marleen van Rijswick and others, 'Ten Building Blocks for Sustainable Water Governance: An Integrated Method to Assess the Governance of Water' [2014] Water International 1 <www.tandfonline.com/doi/abs/10.1080/02508060.2014.951828> accessed 8 September 2014. p. 5.

1016 Food and Agriculture Organization of the United Nations, *Land Resources Information Systems in the Caribbean* (2001).

1017 Algemeen Bureau voor de Statistiek – Suriname, Conservation International and UNDP.

1018 See Ministry of Labour Technological Development and Environment, 'Second National Communication to the United Nations Framework Convention on Climate Change' (2013) <www.nimos.org/smartcms/downloads/1. SURINAME SNC-UNFC on ClimateChange February 2013.pdf>. p. 28.

1019 "Water for People, Water for Life" United Nations World Water Development Report, UNESCO, 2003; Outcome workshop.
See Algemeen Bureau voor de Statistiek – Suriname, Conservation International and UNDP for the water consumption of those they supply to.

experts[1020] that the groundwater resources and thereby the public water supply are under threat. Firstly, saltwater intrusion has been taking place due to over-extraction, especially in the upper north of the coastal area.[1021] Secondly, the sand layer protecting the aquifer in places like Republiek is highly permeable making it challenging to protect it from pollution.[1022] Moreover, it is estimated that by 2050 the production of water will have decreased, thereby reducing the availability of water per inhabitant by 60 percent compared to the current availability.[1023] As a result, the Government of Suriname and SWM, the state-owned company which is responsible for the public water supply, are looking for other sources of safe water which can be sustainably used such as surface water.[1024]

This company has formulated its policy in the Suriname Water Supply Master Plan (SWSMP) which will be rolled out over a period of 14 years (2010–2024).[1025] Because of the Masterplan it can be said that the SWM has a good knowledge of the water system. It could however gain more knowledge about the effects that climate change will expectedly have on freshwater resources. Also, as the SWM might need to make use of surface water in the future it is recommended that it should anticipate this situation and start experimenting with appropriate systems. The SWM could discuss this with the relevant ministries. Keeping the water-ways clean will allow communities to meet their own needs with appropriate measures. It can also be advised for SWM to substitute parts of the infrastructure in Paramaribo. Nevertheless, the SWM currently has a clear understanding of which parts need to be replaced.

Brokopondo

Access to water in Brokopondo largely depends on the quality of surface water which is threatened by different forms of pollution such as mercury due to small-scale gold mining. In this area the Suriname River is the largest river and many tribal and indigenous communities have settled alongside it.

Water resources in these areas are not only contaminated with mercury but with (human) waste. Many of the communities in Brokopondo do not have access to waste management services and as a result the water ways are being used to dispose of waste including human waste as access to sanitation is also limited. Often latrines and sceptic

1020 See: Thomas C Webster and Laura Roebuck, 'Water Resources Assessment of Suriname' (2001).

1021 Government of Suriname and Inter-American Development Bank, 'Suriname Water Supply Infrastructure Rehabilitation' (2010). p. 4.
Interview with Alwin Linger (SWM).

1022 Interview with Donnevan Eduards (SWM).

1023 Nancy del Prado, 'Naar Geintegreerd Waterbeheer in Suriname' (2011). p. 1.

1024 Interview with Alwin Linger (SWM)
https://www.waterkant.net/suriname/2014/05/01/suriname-naarstig-op-zoek-naar-drinkwater/

1025 N.V. Surinaamse Waterleiding Maatschappij, 'Suriname Water Supply Master Plan'.

tanks are emptied into the river. Also, the water in most creeks and rivers is extremely turbid, making it challenging for communities to fish.[1026]

Water in Brokopondo is also used for hydropower, and electricity generated from the reservoir is the main source of power for the country. The Afobaka dam was built to support the power needs of an aluminium smelter. The construction of the dam commenced in 1961 and was completed in 1964. Around 5,000 people, mostly Maroon people, were displaced to build the dam.

What should be established for the district of Brokopondo is the current level of pollution. The state could decide to collect data in a participatory manner as some of the areas might not be easily accessible and it would assist local communities in developing an understanding of the water issues in their region and the impact of their actions.[1027] Otherwise it could delegate government agencies to be responsible for all levels of water screening.

Even though there is a general perception that Suriname has an abundance of water resources, this building block demonstrates that fresh water resources are being challenged due to sea water intrusion and pollution and the public water supply depends on these resources. Therefore, it is important that the quality of these water resources is monitored and protected. As mentioned in the previous chapter the realisation of the human right to water is highly dependent on and interlinked with the quality of the water resources and the environment in general.

7.5.2 BUILDING BLOCK 2: RESPONSIBILITY, AUTHORITY AND CAPACITY

Assessment criteria: Which parties have which responsibilities in realizing the human right to water for vulnerable people? Are the responsibilities clear as well as their responding mandates? Are contracts between States and service providers fully compliant with human rights and sustainability standards? Are there capacity-building strategies for stakeholders?

National level

According to international human rights law the Government of Suriname has the obligation to respect, protect and fulfil the human right to water for those within its territory. It should pay special attention to the needs of vulnerable people. States may choose to delegate tasks to for instance private parties.[1028] However, the final

1026 Heemskerk and Kooye. p. 10.
1027 A similar approach was taking concerning the REDD+ Agenda Ruben Ravenberg, 'Water Management in Suriname' (2013). p. 13.
1028 See chapter 3.

responsibility will remain with the government. In addition, the government is required to protect individuals from infringements of their human right to water by third parties by for instance implementing effective remedies.

Figure 7.2 Overview of relevant governmental agencies and ministries responsible for the public water supply in Paramaribo

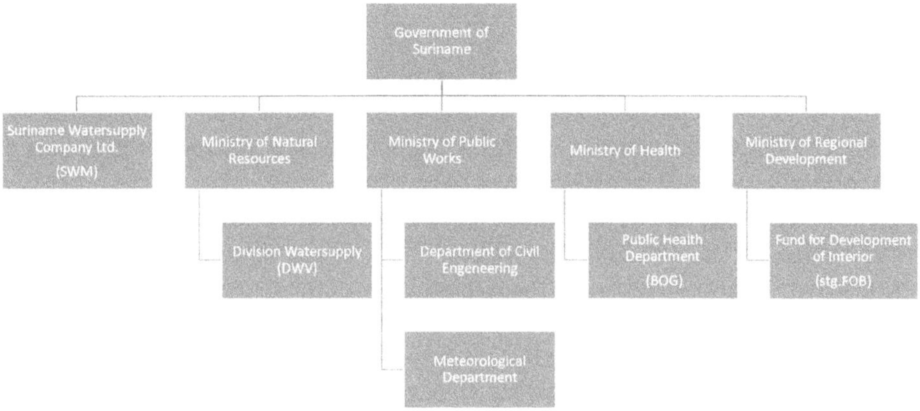

Paramaribo

The N.V. Surinaamsche Waterleiding Maatschappij (SWM), a government-owned limited liability company, is responsible for the public water supply in Paramaribo and other urban areas along the coast.[1029] In these areas coverage is between 65 and 92,6 percent. The Dienst Watervoorziening (DWV) and the Fund for the Development of the Interior (FOB) are responsible for the remaining areas. The government granted a concession contract to SWM in 1933 for a period of 50 years, in which it was allowed to abstract, treat and distribute water in Paramaribo.[1030] This concession expired in 1982. The Government has nevertheless not amended or extended this concession but it did assume that SWM would continue its activities. Finally, a judgement state that as long as the Government does not state otherwise, it will be assumed that the expiration of the concession will not be effectuated.[1031] As a result, the agreements concerning extraction find their basis in a judicial decision. In order to gain an understanding of other agreements between the government and SWM the statutes of SWM[1032] and policy documents which elaborate on the responsibilities of SWM provide additional information.

The SWM has several tasks:

1029 Http://gfcnieuws.org/regering-wil-een-waterleidingbedrijf-voor-gehele-kustvlakte/

1030 N.V. Surinaamse Waterleiding Maatschappij, 'Suriname Water Supply Master Plan'. p. 49.

1031 Interview Gloria Sandie & Alwin Linger (SWM)
Ministerie van Koloniën, Gouvernement van Suriname (1944) 'Concessieverordening G.B. No. 129'

1032 Ministerie van Justitie en Politie, Overheid Suriname (1979) Statuten van de N.V. Surinaamsche Waterleiding Maatschappij (SWM).

a) Where the infrastructure is suitable/ appropriate SWM is assigned to place in-house connections by connecting households to the public water network and where the infrastructure is lacking SWM is required to place public standpipes and taps in yards.[1033]
b) To maintain the water infrastructure in areas appointed by the government
c) To extract groundwater and make it sufficiently suitable to be distributed as drinking water through the water network.[1034]

Even though, SWM is state-owned and the government is the only shareholder, it is still considered to be a company and therefore does not have the same responsibilities as a state.[1035] Companies are however expected to obey the law and respect principles of international instruments, such as human rights.[1036]

The human right to water includes requirements concerning availability, quality, acceptability, accessibility and affordability.[1037] Assessing whether the SWM meets the necessary quality standards is the responsibility of the Public Health Department (BOG). This mandate does not have a legal basis, however. In addition, due to a lack of resources little water quality control actually takes place.[1038] In response, SWM has assumed responsibility for water quality monitoring by using the standards set by the WHO.[1039] The requirement of affordability is controlled by the government. In 2015, for instance, SWM requested that the tariffs be adjusted. The Minister of Natural Resources presented these ideas to Parliament which had to vote on the matter.[1040] In general, the plans of the SWM have to be approved by the government while the Minister of Natural Resources monitors the actions of SWM and is responsible for communications between the government, Parliament and SWM. Also, SWM budget is part of the overall budget of the Ministry of Natural Resources.[1041]

Brokopondo

Other agencies are responsible for providing access to water for domestic purposes. In Brokopondo only between 11 and 20 percent of the population have access to safe drinking water.[1042] As explained, many individuals and communities depend on water from rivers, creeks or rainwater collected in cisterns. These are not considered to be safe

1033 ???
1034 Www.dna.sr/media/141431/SB_2016_no_83_NH.pdf p. 49.
1035 Chapter 3.
1036 United Nations, 'Guiding Principles on Business and Human Rights Guiding Principles on Business and Human Rights' 35. p.1.
1037 Committee on Economic Social and Cultural Rights.
1038 PAHO/WHO. p. 17 & N.V. Surinaamse Waterleiding Maatschappij, 'Suriname Water Supply Master Plan', p. 50.
1039 Interview with Alwin Linger (SWM).
1040 Www.de-surinaamse-krant.com/2015/09/swm-stuurt-nieuwe-tariefvoorstellen.html.
1041 See for instance www.dna.sr/media/45213/S.B._2011_no._118_begr._min_van_NH.pdf.
1042 See the section on the situation of the human right to water in Suriname.

as they are vulnerable to pollution by (human) waste as Brokopondo does not have a landfill site[1043] and harmful chemicals are present, such as mercury. In order to provide access to safe drinking water, the government should facilitate access to safe sources.

Figure 7.3 Overview relevant governmental agencies and ministries responsible for public water supply in Brokopondo

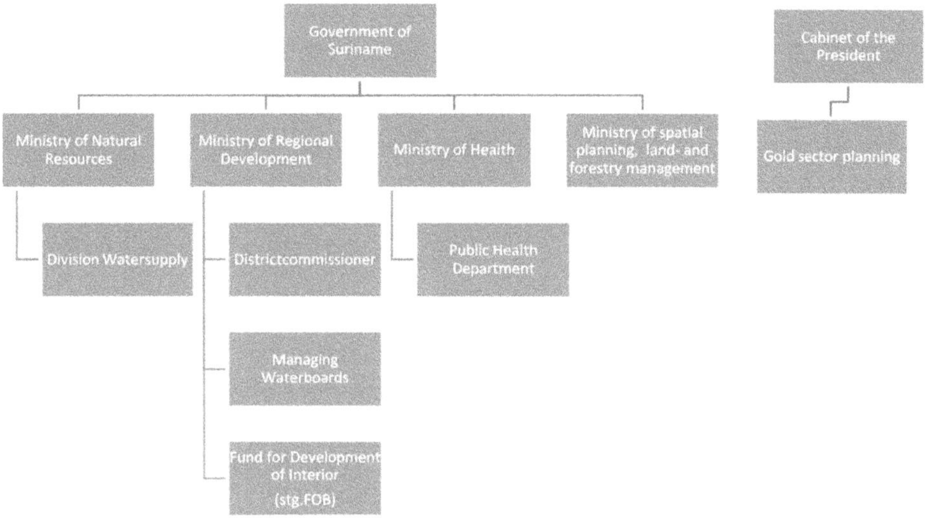

The competent authorities in this region are the Water Supply Division of the Ministry of Natural Resources (DWV) and the Fund for the Development of the Interior (FOB). These areas do not have a sufficient infrastructure to support the public water supply as they are sparsely populated and contain great swathes of untouched Amazon rainforest. Providing water in these areas is by means of water tanks and installing rain-harvesting systems.[1044] Also, NGOs such as the Foundation for the Development of the Interior[1045] and the National Women's Movement as well as private actors have contributed to providing access in the Interior by, for instance, excavating wells.[1046] However, these NGOs are not required to do this under international human rights law as they are formally not responsible for this task. Efforts are being made to provide access to water for at least 50 percent of those living in the Interior, including Brokopondo.

Besides providing access to water for domestic purposes by transporting water to households, ensuring that waterways are clean enough for individuals and communities to facilitate their own access to water is also important. The biggest threats to the

1043 Planbureau Suriname, 'Regionaal Plan Brokopondo 2015 – 2040' (2013), p. 14.
1044 Www.gov.sr/ministerie-van-nh/actueel/nh-verbetert-dienstverlening-met-nieuwe-mobiele-watertank-1.aspx.
1045 PAHO/WHO, p. 17.
1046 Interview with Cheryl Sastro (National Women's Movement).

quality of water in rivers are mercury from small-scale gold mining and other extractive industries and human waste. In relation to this the Ministry of Health is concerned with environmental hygiene and it therefore inspects housing conditions such as access to sanitation and the quality of the living environment. The Ministry of Health also monitors the health of residents in the Interior. Therefore, this governmental organisation should be aware of, for instance, an increase in neurological complaints but also the common occurrence of diarrhoea due to a lack of safe drinking water.[1047] The Ministry of Regional Development is responsible for constructing sanitation and waste management facilities in the Interior.

Overseeing the gold mining sector, including small-scale gold mining, is the task of the Commission for the Planning of the Gold Mining Sector (OGS). This governmental agency oversees the exploitation and extraction of gold by approving licences and monitors its optimal use. In addition, it enforces relevant legislation such as the Mining Law. The Ministry of Land and Forestry Management cooperates with the OGS when it assigns plots of lands for gold exploration. This organisation can contribute to the protection of waterways from pollution by mercury from small-scale gold mining and cyanide by large extractive companies.

One other threat to access to drinking water for those living in Brokopondo are the floods which have been increasingly taking place. They result in the contamination of wells and traditional water sources. Suriname does not have a national natural disaster response force. Task forces are instead established on an ad hoc basis. In May 2006, serious flooding took place along the Suriname River which is the main river in Brokopondo, resulting in the deaths of several individuals. The Pan American Health Organization (PAHO) and the World Health Organisation (WHO) assisted in alleviating some of the adverse effects by donating rain water tanks and purification tablets. In the future, it is expected that natural disasters and floods will become reoccurring events as a result of climate change. In other areas of Suriname recent floods have also taken place.

The responsibilities for providing access to water, as divided between, on the one hand water services provided through the closed water supply system and, on the other, water access by other means, allow the responsible authorities to focus on the particular needs and measures needed. In this case, the separation of tasks can have a positive effect on the provision of water services. However, tasks relating to the protection of water resources on which the right to water for domestic purposes depends are also divided among different government agencies. This requires careful cooperation. Currently, in practice it seems that the tasks and competences are not always clear. As a result, gaps occur where none of the relevant agencies take responsibility.[1048]

1047 Ministry of Labour, Technological Development and Environment. p. 87.

1048 Result of the workshop organized by UNDP, Ministry of Natural Resources and Waterforum Suriname.

7.5.3 BUILDING BLOCK 3: STAKEHOLDER PARTICIPATION

Assessment criteria: Are vulnerable groups identified and involved in the water management which affects them? Are vulnerable groups freely, actively and meaningfully involved? How are the interests, concerns and values balanced in the problem analysis and the decision-making concerning the design, implementation and monitoring of the solution?

This building block covers both active, free and meaningful participation as well as access to information to support the right to participation as explained in chapter 5. Access to information complement the right to participation as it contributes to empowering vulnerable groups and countering unequal power dimensions that affect these vulnerable groups.[1049] It also assists in holding relevant authorities to account.[1050] The requirement of access to justice which is often also related to the right to participation will be discussed in the building block concerning conflict prevention and resolution. Vulnerable individuals should have the opportunity to engage meaningfully in all aspects related to their access to water and sanitation.[1051] In this respect, the voice of vulnerable groups is extremely important as it is often not taken into account when interests are weighed and the principle of non-discrimination is in such cases disregarded.

National level

Both electoral and political participation are relatively high in Suriname, although civic participation can be qualified as weak. The International Development Bank (IDB) explains that this is caused by the historic dependence on the Netherlands and economic and political factors.[1052] If individuals want to influence policy or wish to change something on an individual basis, they will often make use of the personal relationships that they have.[1053]

1049 Catarina De Albuquerque and Virginia Roaf, 'Principles', *Realising the human rights to water and sanitation: A handbook by the UN Special Rapporteur Catarina De Albuquerque* (2014). p. 35.

1050 Catarina De Albuquerque and Virginia Roaf, *On The Right Track. Good Practices in Realising the Rights to Water and Sanitation* (2012th edn, 2012) <www.ohchr.org/Documents/Issues/Water/BookonGoodPractices_en.pdf>. p. 31.

1051 Ibid. p. 33.

1052 International Development Bank, 'Governance in Suriname' (2001). p. 164.

1053 For instance resources are often divided along ethnic lines and political associations. See Marianne Bruin, 'Erken Ons VIDS En OIS, de Stem van Inheemsen in Suriname' (Utrecht University 2008). p. 37.

Figure 7.4 Overview organizational structure District Councils

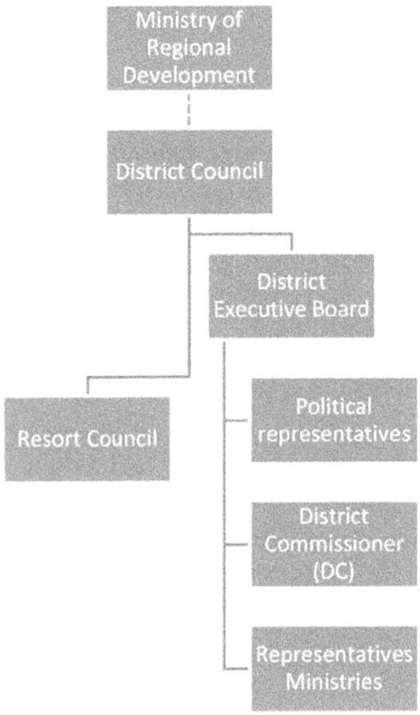

Suriname has been making efforts to improve democratic governance and the participation of citizens by aiming for decentralisation.[1054] For this reason Suriname was divided in 10 districts in 1987: Brokopondo, Commewijne, Coronie, Marowijne, Para, Nickerie, Paramaribo, Saramacca, Sipaliwini and Wanica.[1055] At the head of each district is the District Commissioner (DC) who has executive responsibilities and is appointed by the central government. The DC and the representatives of ministries operating in the district make up the daily management of the district. Each district also has a Council[1056] which consists of political representatives, chosen during general elections.[1057]

The 10 districts of Suriname have been divided into smaller units called resorts with corresponding 'resort councils' to make the management of the areas more effective and to make participation accessible to the local community.[1058] These resort councils[1059] have limited monitoring and executive competences. The councils are supposed to

1054 'Handleiding Burgerparticipatie En Communicatie Voor de Regionale Organen' <http://basahmadali.nl/wp-content/uploads/2015/04/Handleiding-Burgerparticipatie-en-Communicatie-Districtsraad-Paramaribo-06-07-2012.pdf>. p. 3.

1055 Wet Regionale Organen (SB 1989, no. 44)

1056 See Wet Regionale Organen.

1057 Article 162 Constitution.

1058 The resort councils are elected on the same day as the Parliament during general elections which take place once every 5 years.

1059 Article 163 Constitution.

signal problems which occur within the resort and to communicate this to the DC and the central government seated in Paramaribo. In the case of Paramaribo, the district councils and resort councils already have relatively uncomplicated access to the national government to communicate their concerns.

Paramaribo

As SWM provides access to water for domestic purposes in Paramaribo, individuals should have means to participate and have access to information. SWM provides information about its services through infomercials on TV[1060], radio broadcasts and by means of bill-boards scattered throughout the city. Nevertheless, users do not receive personalised information about their water supply. The users in Sunny Point, for instance, indicate that they do not receive any information at all about their water supply from the relevant agencies, the Suriname Water Supply Company Ltd. (SWM) and Division Water supply (DWV).[1061] SWM should make information about its services more easily available. It could do so by posting information on its website or by allowing users to gain more in-depth information at the customer service centres. Not only information about the water services themselves should be made available but also information about the water resources which are used to provide these services.

When it comes to the participation of stakeholders there are very limited opportunities for them to voice their concerns and opinions. Stakeholders are mostly treated as consumers. A distinction can be made between the participation of those who are currently connected to the public water supply and served by SWM and those who are not. Those who are connected to the public water supply have access to the customer service centres and the Interference Department at SWM.[1062] Here they can file complaints and discuss their concerns. Depending on the content of the complaint, it will be addressed by the relevant department.

Those who are currently unserved have limited means to participate. As they are not users, they are not covered by communication policies of SWM. Their right to participate often not respected because they are not recognised by the government as they for instance live in informal settlements and therefore lack property rights. Nevertheless, through political forums certain informal settlements have been successful and have been given commitments for the extension of infrastructure.[1063] This shows that vulnerable groups can be involved in discussing their water supply and other issues when political party meetings take place in their (unserved)

1060 Info Act is a TV program commissioned by the Government which discusses current events. In the episode of 26 February 2016, for instance, the public were informed that 15 kilometers of a water pipeline which contained asbestos had been replaced by PVC pipes near the village of Peperpot, Commewijne.

1061 Sandie. p. 36.

1062 N.V. Surinaamse Waterleiding Maatschappij, 'Suriname Water Supply Master Plan'. p. 58.

1063 See for instance https://www.waterkant.net/suriname/2015/04/28/schoon-drinkwater-voor-krakers/

neighbourhoods. Stakeholders can use these opportunities to share their concerns about their water supply.[1064]

The participation of vulnerable individuals and the public in general seems to be mostly problem based. In the majority of cases individuals approach the SWM to discuss their concerns when there is a problem. However, according to the human rights framework communities and individuals should have the right to determine the type of sanitation and water services they need and have a say in how these services are managed. They may even decide to manage their own services with assistance from the State.[1065] According to the former Special Rapporteur:

> *National and local standards and targets should therefore be based on studies and consultations to ascertain what different individuals or groups, particularly those who are disadvantaged, consider to be the most essential aspects of their water and sanitation services, so as to ensure the relevance of the standards to the local context.*[1066]

The SWM should therefore establish a structural system which allows stakeholders to participate meaningfully and not only when this is problem based. Vulnerable individuals should be engaged during the design, implementation and monitoring of the system. Consumers should be made aware of which possibilities for participation are available to them. The SWM and the government of Suriname can take other states which have successfully implemented a participatory process in their water management policy as examples.

What is often seen as a challenge is the lack of property rights by certain vulnerable individuals. Even if individuals and groups are illegally occupying settlements, these communities still retain their human right to water and their right to participate. Property status is a prohibited ground for discrimination.[1067] Access to water services and the right to participation should not therefore be made conditional upon their land tenure status as is currently the practice in Suriname. Even worse, these communities of informal settlements have also experienced the forced removal of their communities and the destruction of their improvised homes.[1068] Their housing situation should be respected and improved or, alternatively, housing which does meet human rights standards should be provided. As a part of this, these communities should structurally be involved in the decision-making processes for the services that they will receive. States must find solutions which are appropriate. They may choose to provide access to water by installing a public water tap which is accessible outside on a public plot of land so that illegal occupants in the neighbourhood can access water there. It should again be emphasised that in designing

---

1064 This can lead to substantive improvements such as was the case in Maystraat and Rodekruislaan, Paramaribo. See https://www.waterkant.net/suriname/2015/04/28/schoon-drinkwater-voor-krakers/

1065 Catarina De Albuquerque and Virginia Roaf, 'Services', *Realising the human rights to water and sanitation: A handbook by the UN Special Rapporteur Catarina De Albuquerque* (2014), p. 14.

1066 Ibid. p. 14.

1067 International Covenant on Economic, Social and Cultural Rights 2009 1, paragraph 25.

1068 See Viyantie Ramlal, 'De Huisvestingsproblematiek in Suriname' (Anton de Kom Universiteit van Suriname 2013).

the most appropriate service, the public in general and vulnerable individuals should be included. Having these groups actively participate also enhances the sustainability of the water services and empowers vulnerable groups such as women.[1069]

Participation in relation to the SWM is not the only relevant means of participation in relation to water services. If vulnerable groups want to influence the overall policy concerning the development of their area in Paramaribo they have access to district and resort councils where, among other things, the allocation of financial resources is discussed.

Brokopondo

In the case of Brokopondo, participation in relation to water access can primarily take place through the District and Resort Councils. This is because their water access is the responsibility of several governmental agencies, specifically the DWV. As mentioned, it is estimated that between 11 and 12 percent of those living in Brokopondo have access to safe drinking water. Stakeholder participation should provide opportunities for those currently unserved to express their opinions and be involved in the decision-making processes which affect their access to safe water. In Brokopondo there are two major issues: not receiving water services which meet the standards of the human right to water and the pollution of water resources that communities use to meet their water needs. In order to tackle these two issues stakeholder participation by vulnerable groups is essential.

Citizens can voice their opinions in several ways in the Interior. Firstly, by electing those who they are convinced will effectively represent their interests at the local level. And secondly, by participating in resort meetings where, for instance, the budget of the resort or district is decided upon. The District is obligated to involve citizens in the development of district and resort plans according to Article 50 of the Regional Development Act (Wet Regionale Ontwikkeling (WRO). These plans are however still subject to approval by the central government, which upon approval designates financial resources to the districts and resorts.[1070]

As mentioned, according to the human rights framework participation needs to be meaningful. When stakeholders share their views at these council meetings, they should be taken into account. Stakeholders should be able to recognise where and how their opinions are considered. In practice, the current system of local participation is not being properly executed and does not therefore effectively assist in the realisation of the human right to water for those who are currently unserved.[1071] Stakeholder meetings often do not take place at all or not on a regular basis. This has led to stakeholders withdrawing from the participation process as they question the effectiveness thereof.[1072]

1069 De Albuquerque and Roaf, *On The Right Track. Good Practices in Realising the Rights to Water and Sanitation*. p. 29.

1070 Interview with Thanya Soke-Fonkel (Ministry of Regional Development).

1071 Kambel and MacKay. p. 161.

1072 Www.srherald.com/suriname/2017/02/16/participatie-burgers-para-aan-hoorzittingen-niet-optimaal/

The participation of stakeholders in general is an area which needs to be developed in Suriname. However, when it comes to the participation of groups such as tribal and maroon peoples specifically, the government has been criticised for disregarding their rights to participation, consultation and information, especially when this is related to assigning concessions to extractive industries to exploit natural resources in Brokopondo.

For instance, in 1990 the central government allowed several multinational logging and mining companies to undertake activities on land where the Saramaka people were located. The Saramaka people had not been consulted even though they had been living there for many generations. The government had taken the position that since the Constitution specifies that all non-titled land and resources belong to the State, the State did not need to include the Saramaka in the decision-making process. In 2000, the Saramaka, the largest maroon tribe in Suriname, petitioned the Inter-American Commission for Human Rights which led to a landmark decision by the Inter-American Court, 17 years after the beginning of the dispute, in 2007. The Court demanded that states have certain procedural safeguards in place when issuing concessions for natural resources. In relation to this it made the following statement:

> *'[T]he State must abide by the following three safeguards: First, the State must ensure the effective participation of the members of the Saramaka people, in conformity with their customs and traditions, regarding any development, investment, exploration or extraction plan (hereinafter "development or investment plan")127 within Saramaka territory. Second, the State must guarantee that the Saramakas will receive a reasonable benefit from any such plan within their territory. Thirdly, the State must ensure that no concession will be issued within Saramaka territory unless and until independent and technically capable entities, with the State's supervision, perform a prior environmental and social impact assessment.'*[1073]

In the following paragraph the Court continued to elaborate on the right of participation explaining that indigenous and tribal peoples must be consulted in good faith, in a culturally appropriate manner and should have access to sufficient information. The Court explained where it found its reasoning for these requirements in relation to participation by indigenous and tribal peoples.

> *'First, the Court has stated that in ensuring the effective participation of members of the Saramaka people in development or investment plans within their territory, the State has a duty to actively consult with said community according to their customs and traditions (supra para. 129). This duty requires the State to both accept and disseminate information, and entails constant communication between the parties. These consultations must be in good faith, through culturally appropriate procedures and with the objective of reaching an agreement. Furthermore, the Saramakas must be consulted, in accordance with their own traditions, at the early stages of a development or investment plan, not only when the need arises to obtain approval from the community, if such is the case. Early notice provides time for internal*

[1073] Case of the Saramaka People v Suriname, Interpretation of the judgment on preliminary objections, merits, reparations and costs, IACHR Series C No 185, IHRL 3058 paragraph 129.

discussion within communities and for proper feedback to the State. The State must also ensure that members of the Saramaka people are aware of possible risks, including environmental and health risks, in order that the proposed development or investment plan is accepted knowingly and voluntarily. Finally, consultation should take account of the Saramaka people's traditional methods of decision-making.'[1074]

The Court decided that Suriname had not established the necessary safeguards when it granted the concessions on the traditional tribal lands of the Saramaka people. The judgement has so far not been implemented. And, in 2015, another judgement was delivered against Suriname relating to the participation of, in this instance, the Kalina and Lokono indigenous people. The Court repeated the arguments it had made in previous judgements by finding 'the absence of effective participation, by means of a consultation process, with regard to the mining concessions within one of the nature reserves in the ancestral territory.'[1075]

Concerning these cases the Court adopted a series of orders with corresponding deadlines. The Surinamese government has thus far not amended its legislation or met the other requirements mentioned by the Court. Therefore, meaningful participation by indigenous and tribal communities has still not been implemented. Suriname should adopt the measures which were advised by the Inter-American Court. In addition to a national plan for citizen and stakeholder participation should be developed with special attention being given to the participation of vulnerable groups.

The sections above have demonstrated that participation in relation to water services and water management should be further developed to ensure the participation of vulnerable individuals as this is currently fragmented. A way in which participation in relation to water management could be reformed and enhanced is by introducing Water Boards in Brokopondo. As Water Board have participation of stakeholders as part of their framework.[1076] Suriname already has experience with Water Boards. In order to manage where and how water is allocated, and to ensure that food production is sufficient, five Water Boards were established in 1931, following the Dutch example, in Nickerie, West Suriname. The Surinamese regulatory system also allows for Water Boards to be established in other regions. For instance, in 2008 the idea to establish a Water Board for the management of the Para River was presented.[1077] Currently, the Ministry of Regional Development is contemplating whether a Water Board and managing the Suriname River in its catchment area could be beneficial. For the participation of vulnerable groups and the inclusion of all relevant stakeholders which are relevant for tackling issues such as water pollution, installing Water Boards could be a valuable asset. Water Boards can give an equal voice to all relevant local stakeholders if they are set up in the appropriate

---

1074 Case of the Saramaka People v. Suriname paragraph 133.

1075 Paragraph 127.

1076 See Heleen Kole, 'Het Poldermodel. Politieke Cultuur in de Waterschappen (c.1400-c.1800).' (Utrecht University 2017).

1077 Www.planningofficesuriname.com/wp-content/uploads/2014/12/Waterbeheersingsplan-Pararivier-en-Zijkreken.pdf.

manner, for instance by granting voting rights. Water Boards can allow for local issues to be taking into account. Local communities can therefore contribute to problem solving.

Participation is a key element if vulnerable groups' human right to water of is to be realised. This building block shows that the participation of consumers is not structured in an accessible and proactive manner. Consumers are often only included when financial issues are at stake and sometimes when they instigate a complaint procedure concerning the quality of the water that they receive. However, the human rights framework requires, for instance, that individuals are able to express their views about the water supply system on which they depend. In addition, in order to form a well-informed opinion access to information should be enhanced and made easier. In Brokopondo, the participation of individuals has to go through several stages which tend to dilute the opinions of those concerned. For this reason, individuals are not aware of the extent to which their opinions actually influence policy making. Also, their ideas on how financial resources should be allocated are often not sought. In general, the participation of those living in the Interior is often disregarded as the court cases demonstrate. This is often the case because the interests of those in the Interior clash with the ambitions of those in the city.

7.5.4 BUILDING BLOCK 4: VALUES, PRINCIPLES AND POLICY DISCOURSES

Assessment criteria: Is there sufficient knowledge of shared or conflicting values, viewpoints and principles (represented by different policy discourse coalitions) for water issues and their consequences for facing water management issues?

National level

At the national level an important instrument which reflects the values and principles which are part of Surinamese policy, is the Development Plan. This instrument provides an overview of the goals which the Government has set for itself. It was adopted by a majority vote in Parliament. The Development Plan 2012 – 2016 is the most recent version as the Development Plan for the forthcoming four years has not yet been approved.[1078] However, the draft Development Plan 2017–2024 also includes a section on water management which mentions improving the public water supply infrastructure as a priority and the adoption of four water laws.[1079] The Development Plan 2012 – 2016, which is still in effect, establishes certain priority issues: waste management, renewable energy, protection of the atmosphere, sustainable water management, and environmental and sustainable land management.[1080] The Development Plan uses the

1078 Interview with Revinh Ramnandanlal (Cabinet of the Vice President).
1079 Paragraph V. 3.2.
1080 Regering van de Republiek Suriname, 'Ontwikkelingsplan 2012 – 2016 Suriname in Transformatie' (2012).

norms and requirements of the Convention on Biodiversity for the formulation of these goals. Access to water is also mentioned in this document as an important social goal.

According to the current Development Plan: "Development focuses on people. Everything depends on humans first. All individuals, from the cradle to the grave, need for a minimum level of living, will take centre stage and thus have priority in management and policy. A general basic standard of living, including the availability of water, and power for every household, access to education for all and accessible, attainable feasibility and affordability of health care for all Surinamese, will be indicated. Development is intended to enhance the living standard."[1081] In line with this thought the Surinamese government aims to meet the Sustainable Development Goals (SDGs). It has determined water resources management as the third priority but it does not elaborate on how it will reach the goals which relate to providing access to water for domestic purposes.[1082]

Providing access to water to all citizens is considered an important social goal by the Surinamese government. However, it has not explicitly recognised a human right to water in its Constitution nor has it explicitly supported the right at international level. The government does however, discuss the importance of access to water in other documents like the Development Plan.

This instrument explains about the sustainability strategy of the Surinamese government that[1083]:

> *"The Surinamese economy is "two-legged": the unsustainable industries and the sustainable sectors. The unsustainable sectors, on the one hand, such as bauxite, gold, and oil are finite resources and produce high revenues. These sectors, which are among the non-renewable resources, will have to provide capital for the development of the renewable sectors: agriculture, forestry, public services, tourism, as well as the social, educational, medical and cultural sectors."*[1084]

Paramaribo

It is the aim of SWM to assure 24/7 service and increase the coverage in the coastal area to 95 percent.[1085] That access to water for domestic purposes is a shared value is expressed in the relatively low cost of water services in Paramaribo. Consumers' water tariffs for water for domestic purposes do not cover production costs.[1086] It seems that the government has adopted a solidarity approach by having used public funds in the past to cover part of the costs. This notion of solidarity can also be recognised when looking at the water services for the Interior. In certain instances, these water services are free and the costs are covered by public funds.[1087]

1081 Ibid. p. 37 Translated by author.
1082 Government of the Republic of Suriname, 'National Report in Preparation of The Third International Conference on Small Island Developing States (SIDS)' (2013). p. 4.
1083 Translated by author.
1084 Regering van de Republiek Suriname. p. 38.
1085 N.V. Surinaamse Waterleiding Maatschappij, 'Suriname Water Supply Master Plan'. p. 44.
1086 See financial arrangements.
1087 For instance in Saramacca.

It seems to be a shared value that everyone should have access to safe drinking water.[1088] The reason to why this goal is not currently being met possibly lie in the trade-offs that are being made. In order to circumvent other interests from undermining this goal, access to water for domestic purposes should be prioritised in legislation and policy. It also seems that access to water is seen as a need but not necessarily as a right for which the government is responsible for its realisation.[1089] There also seems to be a lack of convergence between how water resources are used and which effect this has on the drinking water supply.

Brokopondo

Providing access to water in Brokopondo, not only relates to the overall ideas the government has about providing access but also the values and principles which apply to the resource itself. Central provisions which represent the values of the Surinamese government relating to access to water and water management can be found in the Constitution. There is one provision which serves as an umbrella provision for social, economic and environmental development even though these might conflict with each other when translated into specific regulations. '[t]he social objectives of the State shall aim at (...) [c]reating and improving the conditions necessary for the protection of nature and for the preservation of the ecological balance.'[1090] This provision though general in its wording, could prove influential in sustainable and equitable water management and a sustainable right to water in the Interior as it is highly dependent on the environment. On the basis of this provision Suriname participates in environmental conventions.[1091]

The Government following the colonial tradition considers itself to be the owner of all natural resources in Suriname.

> *'Natural riches and resources are the property of the nation and shall be used to promote economic, social and cultural development. The nation has the inalienable right to take complete possession of its natural resources in order to utilize them to the benefit of the economic, social and cultural development of Suriname.'*[1092]

This point of view has been contradicted by human rights lawyers who argue that according to international conventions ratified by Suriname, indigenous and tribal peoples have rights to land resources.[1093] In relation to this it is often criticised that the government focuses on the development of the coastal areas and neglects the Interior. As examples the Koffiekamp and Afobaka conflicts are often mentioned. In both situations, tribal communities had to respectively make way for the establishment of mining companies and a dam.[1094] They were told that they would benefit from these developments. However,

1088 Workshop.
1089 Interview Jeugdjournaal.
1090 Article 6(g)
1091 The Ministry of Labour Technological Development and Environment. p. 34.
1092 Art. 41.
1093 See for instance Kambel and MacKay.
1094 Heemskerk and Kooye. p. 7.

these communities are still left without general public services, including electricity which should have been provided for by the Afobaka dam and its water.[1095]

The Government of Suriname has expressed its commitment to sustainable water management and it aims to use the concept of Integrated Water Resources Management to attain this.[1096] In addition, several Surinamese governmental agencies like the Ministry of Natural Resources have indicated that their aim is to meet the Sustainable Development Goals.[1097]

For the tribal and indigenous communities, for whom the right to water should be realised, other values are important. For instance, the Lokono 'have a special physical and spiritual relationship with their lands and natural resources. They consider that all the animals, plants, fish, stones, streams and rivers are interconnected living beings that have protective spirits.'[1098] This worldview conflicts with the perspective of the government. Similar to other indigenous and tribal communities around the world, those in Suriname do not consider the state or other parties as owners of the environment.

This conflict between values can be seen in the conflicts and historical tensions between the indigenous and tribal communities which live in the Interior on the one hand and the other ethnic communities who live in the urbanised regions on the other hand.[1099]

The inclusion of other perspectives, values and principles than those of the Surinamese government could provide for more holistic and sustainable points of view. Especially when access to water for tribal people is to be realised, these values should be taken into account. As was explained in chapter 3, water services should be provided in a culturally appropriate manner. Examples of the inclusion of non-Western ideas in water management can for instance be found in the recent developments in New Zealand and India where the perspective of the indigenous Maori and Hindu philosophy respectively has resulted in the recognition of rivers as having legal personality.[1100]

In general, the idea that access to water should be universal is shared as a value in society. Whether it is accepted as a right is debatable. It is most commonly approached as a basic need and not necessarily an entitlement. In the Interior, access to water is dependent on the quality of traditional water resources such as lakes. When it comes to these resources, the values and principles clash as the government considers itself the owner of all resources and those in the Interior consider natural resources to fall outside

---

1095 Ellen Rose Kambel, 'Mijnbouw, Infrastructuur En de Rechten van Inheemse En Tribale Volkeren in Zuid-Oost Suriname' [2009] Project strategische analse en participatief actieplan voor zuid oost Suriname.p. 12.

1096 Regering van de Republiek Suriname. p. 145.

1097 Interview with Wilco Finisie (Ministry of Regional Development)
Result of workshop, statement made by Director of Ministry of Natural Resources.

1098 Kaliña and Lokono peoples vs. Suriname paragraph 35.

1099 Heemskerk and Kooye. p. 7.

1100 See James DK Morris and Jacinta Ruru, 'Giving Voice to Rivers: Legal Personality as a Vehicle for Recognizing Indigenous Peoples' Relationship to Water?' (2011) 14 Australian Indigenous Law Review 49.
Sangeetha Yogendran. (2017). Natural Entities Now Legally People: Enough to Save Them? (RSIS Commentaries, No. 087). RSIS Commentaries. Singapore: Nanyang Technological University.

of any ownership. These understandings and values influence how water recourses in Suriname are protected.

7.5.5 BUILDING BLOCK 5: REGULATIONS AND AGREEMENTS

Assessment criteria: How do regulations and agreements recognise or support a sustainable human right to water? What are the main problems with regard to the above-mentioned aspects?

International level

The Surinamese's legal order adopts a monistic approach to the connection between national and international law. It thereby accepts that the national and the international legal order form a unity. In this system, the international law has a higher status than national law and has preference over national law in the event of a conflict.[1101] Article 105 of the Constitution establishes that treaties 'which may be directly applicable to anyone shall have this binding effect as from the time of publication.[1102]

Table 7.6 Selection international instruments Suriname is party to

Instrument	Entry into force
International Covenant on Civil and Political Rights	1976
Optional Protocol to the International Covenant on Civil and Political Rights	1976
International Convention on the Elimination of All Forms of Racial Discrimination	1984
International Covenant on Economic, Social and Cultural Rights	1977
American Convention on Human Rights	1987
Convention on the Elimination of All Forms of Discrimination against Women	1993
Convention on the Rights of the Child	1993
Convention on Biological Diversity	1996
Convention concerning the Protection of the World Cultural and Natural Heritage	1998
United Nations Framework Convention on Climate Change	1998
Kyoto Protocol to the United Nations Framework Convention on Climate Change	2006
Cartagena Protocol on Biosafety to the Convention on Biological Diversity	2008
Stockholm Convention on Persistent Organic Pollutants	2011
Optional Protocol to the Convention on the Rights of the Child on the involvement of children in armed conflict	2012
Convention on the Rights of Persons with Disabilities	2017

1101 Kambel and MacKay. p. 137.
1102 Constitution article 105.

Suriname has not explicitly recognised the human right to water in its own legislation. However, it is party to the ICESCR, which implicitly recognises this right.[1103] Suriname is also a party to the American Convention on Human Rights and the American Declaration of the Rights and Duties of Man. The Inter-American Commission on Human Rights monitors human rights in all member states and investigates individual petitions alleging violations of the human rights laid down in the American Declaration of the Rights and Duties of Man. If states recognise the competence of the Inter-American Court of Human Rights, these petitions can be referred to this Court. Petitions can be submitted by any persons, groups of persons or non-governmental organisations.

Similar to the UN system, the human right to water has not been explicitly recognised in treaties or declarations of the Inter-American human rights system. The human right to water has been recognised in the resolution on the human right to safe drinking water and sanitation which was adopted in 2012. It invites 'member states, in keeping with their national realities, to continue working to ensure access to safe drinking water and sanitation services for present and future generations.'[1104]

The OAS resolution is less comprehensive than its UNGA counterparts. It seems to focus on the sovereignty of states to exploit their natural resources according to their own policies and refers to Principle 2 of the Rio Declaration.[1105] The OAS does not explicitly state that access to water is a human right, although it does seem to consider it as a derivative right.[1106] There are a limited number of cases in the Inter-American system which concern the human right to water. According to Murillo Chavarro, cases concerning the human right to water either concern conditions in detention or relate to vulnerable groups, especially indigenous peoples.[1107] For instance, in the case concerning the indigenous Yakye Axa community the Court stated the following:

> *'Special detriment to the right to health, and closely tied to this, detriment to the right to food and access to clean water, have a major impact on the right to a decent existence and basic conditions to exercise other human rights, such as the right to education or the right to cultural identity. In the case of indigenous peoples, access to their ancestral lands and to the use and enjoyment of the natural resources found on them is closely linked to obtaining food and access to clean water.'*[1108]

1103 See chapter 3.

1104 Paragraph 1.

1105 Preamble taking note of Principle 2 of the Rio Declaration on Environment and Development, which established that "States have, in accordance with the Charter of the United Nations and the principles of international law, the sovereign right to exploit their own resources pursuant to their own environmental and developmental policies, and the responsibility to ensure that activities within their jurisdiction or control do not cause damage to the environment of other States or of areas beyond the limits of national jurisdiction"

1106 Jimena Murillo Chávarro, *The Human Right to Water A Legal Comparative Perspective at the International, Regional and Domestic Level* (Intersentia 2015). p. 147.

1107 Jimena Murillo Chávarro, 'The Right to Water in the Case-Law of the Inter-American Court of Human Rights' (2014) 7 39. p. 48.

1108 IACtHR, Indigenous Community Yakye Axa v Paraguay, Judgment of June 17, 2005, para. 167.

Based on the aforementioned it can be concluded that within the UN and Inter-American human rights systems Suriname has the obligation to realise the human right to water for its inhabitants. The human rights system leaves it up to the states to decide the exact measures they will take to fulfil this right.

National level

Within national law there are several laws which support the realisation of the human right to water for individuals. These laws are often outdated and lack enforcement, but they still have legal force and could be used. The Water Supply Act, for example, dates back to 1938 and determines restrictions and requirements for access to the public water system in Paramaribo. For instance, Article 1 explains that it is prohibited for individuals to have wells, pits and other facilities to extract groundwater. In addition, individuals are not allowed to have bins and barrels which can be used to save and collect water. Individuals are allowed to apply for a permit to allow a well to be dug and to own items to collect water. However, the permit will only be granted if allowing the wells is in the best interest of the general public.[1109] Unserved communities often collect rainwater in barrels as they cannot depend on ditches which are highly polluted. Also, companies are exempted from this prohibition. A fine of 500 guilders is attached to an infringement.[1110] The following provisions provide an insight into the reasoning behind these restrictions. The restrictions are meant to motivate households to become connected to the public water supply system. Owners of buildings are obliged to make use of this system. The water supply infrastructure was installed in 1933 and the government aimed to promote this closed piped water system.[1111] This legislation only applied to Paramaribo because at the time only this city possessed the necessary infrastructure. The legislation has not been amended since this time even though other areas of Suriname now also have the infrastructure needed to connect them to the public water system.

As mentioned, SWM was granted a concession for extracting ground water. This was based on the Concession Act from 1907 which includes rules which are applicable to the exploitation of groundwater resources for public utilities. The President is authorised to grant concessions for the construction and operation of any public works which are needed. Besides the concessions for public water services, private extraction companies have been on the increase since 2005. Around 14 private companies have been granted concession to bottle water in order to sell it at commercial prices.[1112] This sector is not well regulated and quality standards do not currently apply to these companies. The rise of private water companies, if not coordinated, could encourage the over-extraction of groundwater.[1113]

1109 Article 1(4)(b).
1110 The current currency is the Surinamese Dollar.
1111 Interview with Del Prado (Environmental Law and Policy Consultant).
1112 Surinaamse producenten gebotteld water zwijgzaam over herkomst water – Export drinkwater plan president Bouterse komt niet van de grond http://obsession-magazine.nl/?p=58463.
1113 Interview with Alwin Linger (SWM).

As of April 2017, Suriname does not have regulations and standards specifically relating to water management. In 2007, certain regulations were drafted, but the procedure became stalled and the following regulations are not yet in force as they are currently in the process of being appraised.[1114]

a. Draft act concerning the protection of groundwater withdrawal areas
b. Draft act concerning the withdrawal of groundwater
c. Draft act on the supervision of the water quality of water companies which also includes standards on drinking water quality.

In its Development Plan 2012 – 2016, the Government referred to these regulations and reinforced its intentions.[1115] Nevertheless, at the end of this term the legislation had still not been adopted and implemented or even submitted to Parliament. As a consequence, there are no acts protecting groundwater from overexploitation or pollution. This means that the legal framework is lacking and is unable to control and minimise groundwater withdrawal or to prevent activities which can cause contamination.

Not only are regulations concerning groundwater lacking but also those concerning surface water. This is relevant as surface water will have to be used for the general water supply in the near future. Surface water needs laws that address, amongst other things, the discharging of wastewater and pollution in general.[1116] Although specific water laws are lacking there are also a few environmental policies which are relevant to the water supply such as the Hindrance Act of 1930 which places limitations on activities that can produce any form of hindrance to the general public, including water pollution.[1117] However, the existing legislation addressing health and environmental issues is unclear, vague and outdated. The majority of these Acts are more than 60 years old and therefore the legal framework needs to be updated and expended.[1118]

In addition to the three water acts, there is a draft law establishing a national water authority. This idea emerged in the 1980s, but the legal drafting process was only recently completed and the law will eventually enter into force if Parliament votes in favour.[1119] The Surinamese Water Authority (SWA) will weigh the different interests in the water sector and work towards sustainable water use.[1120] It will monitor and enforce the legal provisions which apply to the water sector.[1121] The SWA is supposed to supervise and enforce the relevant provisions in the three draft water acts mentioned above. The SWA can be asked to advise the Minster of Natural Resources and the

1114 Government of Suriname and Inter-American Development Bank, p. 7.
1115 Regering van de Republiek Suriname, p. 146.
1116 Webster and Roebuck, p. 6.
1117 (Government of Suriname & Inter-American Development Bank, 2010) p. 8.
1118 Ibid. p. 6.
1119 Del Prado, p. 72.
As of April 2017 it is unknown when the Act on the water authority will be voted upon.
1120 Memorie van Toelichting.
1121 Article 4 Wet Surinaamse Water Autoriteit.

Minister of Public Health.[1122] However, its advice will not be binding.[1123] The SWA will also look into general water and environmental issues like floods and wastewater management.[1124] The SWA is supposed to supervise and enforce the relevant provisions in the three draft water acts mentioned above.

The Surinamese government should explicitly recognise the human right to water and prioritise its realisation. The different water laws which have been drafted should be reviewed as their provisions are outdated and vague when it comes to competences and enforcement. This also relates to the SWA. What should be further developed are the specific tasks and competences that this Authority will have. If it is allowed to weigh interests with the overall goal to manage water use sustainably, the Authority should be able to make final decisions in this regard or to issue binding advice.

Besides these draft laws there is also the 2010 draft Environment Act which includes rules concerning conservation, management and protection of a healthy environment. An important aspect of this Act is the requirement of environmental impact assessments for activities which may affect the environment.[1125] The draft law establishes the allowed deposits and escapes of contaminants on land, in soil and water which will be determined during a licencing procedure. It is still uncertain when this law will be approved in Parliament.

The human right to water is not explicitly recognised by Suriname. Nevertheless, the treaties it has signed do create obligations for the government regarding this right. Implementation at the national level and in Paramaribo does not identify all the necessary standards. For this reason, the regulatory framework should be reviewed. Even though this has been done before, the draft laws still demonstrate gaps.

Brokopondo

For the *Brokopondo* district, in addition to the general applicable laws which have been mentioned there are several instruments and policies which apply to water resources management. Starting at the international level, the following instruments are relevant in this regard as they include provisions concerning access to water and the protection of water resources.

In this light, international agreements signed by Suriname which provide individuals with claims to water resources and which create obligations for Suriname concerning these natural resources are directly accessible. This is especially important where the Surinamese legal framework leaves gaps and international law can therefore be used to fill these gaps. These international agreements can lay down certain obligations and standards which facilitate sustainable access to water for vulnerable groups. These include human rights treaties but also agreements concerning nature

[1122] Article 5 and Article 4.
[1123] Article 5 paragraph 2.
[1124] Memorie van Toelichting.
[1125] Del Prado, 'Analysis of Instruments, Principles and Legal Norms in Force on Domestic Law of the Amazon Basin in Water Resources Management'. p. 25.

conservation, biodiversity and the protection of transboundary watercourses such as the Convention on Biological Diversity and the Amazon Cooperation Treaty.

Another instrument which has been signed by Suriname and is relevant for water management is the Treaty for Amazon Cooperation. This instrument was signed by the eight states bordering the Amazon in 1978 and the Amazon Cooperation Treaty Organization (ACTO) was established in 1995 to support the implementation of the Treaty.[1126] The ACTO has as its main focus to achieve 'the preservation of the environment, and the conservation and rational utilisation of the natural resources of those territories.'[1127] The Treaty is legally binding but its content can be said to be relatively soft.[1128] Abi-Saab notes that 'states have large discretion as to what needs to be done and by which means; moreover, their execution can hardly be assessed by an external objective control mechanism, such as a process of judicial control.'[1129] In relation to water this Treaty establishes that member states should 'make efforts aimed at achieving rational utilisation of water resources.'[1130] Therefore it can be said that the Treaty supports the principles of sustainable development such as rational utilisation but also ecological conservation as the preamble states that member parties should 'maintain a balance between economic growth and conservation of the environment.'[1131] The Treaty also mentions that sanitary conditions should be improved to combat epidemic diseases.[1132] Several instruments have been developed; nevertheless, the agreements to prevent water contamination have not been successful.[1133]

The 1930 Nuisance Act which was last amended in 2011 includes provisions and requirements concerning licences granted by the District Commissioner (DC) for activities by, for instance, extractive enterprises. There are special licences for those activities which cause damage. The DC may refuse or revoke a licence when the enterprise causes harm, damages property, has serious negative health effects or damages nature. Therefore, the District Commissioner is authorised to forbid certain activities from taking place if he believes that these activities will cause substantial damage to local water resources. NIMOS is often asked for its advice and it has become common practice for it to provide companies with social and environmental impact assessments on a voluntary basis.

The Constitution does not contain provisions which specifically deal with water. The Constitution does however state that the Republic of Suriname has exclusive rights to the nation's resources. Article 41 of the Constitution of the Republic of Suriname (1987) states the following: 'Natural riches and resources are property of the nation and shall be

---

1126 Article II Amazon Cooperation Treaty.

1127 Article 1.

1128 Beatriz Garcia, 'The 1978 Amazon Cooperation Treaty', *The Amazon from an International Law Perspective*, vol. 1 (Cambridge University Press 2011). p. 95.

1129 Ibid. p. 95.

1130 Article 5.

1131 Preamble.

1132 Article 8.

1133 Ibid. p. 124.
For instance the Amazon Charter.

used to promote economic, social and cultural development. The nation shall have the inalienable right to take complete possession of the natural resources in order to apply them to the needs of the economic, social and cultural development of Suriname.'[1134]

At the national level, there are several provisions which could protect surface water from pollution. The first provision concerns pollution and is included in the Penal Code. In Article 224 of the Penal Code the following is stated: 'He who in a well, pump, water source, trench, creek or in a publicly used designated drinking water facility places any substance, knowing it will make the water harmful to health and life, shall be punished by a term of imprisonment not exceeding fifteen years. If the offence has caused a person's death, the perpetrator shall be punished by life imprisonment or a term of imprisonment not exceeding twenty years.'[1135] This article covers those who have deliberately placed harmful substances in water used for drinking. Article 225 covers those who may not have been directly involved or who may not have deliberately placed harmful substances in drinking water, but who nevertheless bear legal responsibility for this. The question is whether this provision could be applied to the pollution of water resources in Brokopondo. It could provide an avenue to hold small-scale gold miners accountable. However, proving causality would be challenging.

There are several remarks that can be made about these provisions. Firstly, these articles have not been amended since 1911 and 1915 and are therefore in need of updating. One of the problems therewith is that they lack complementary provisions which for instance explain the terms which are used, such as pollution. Secondly, the articles only apply when the harmful effects have actually materialised as a consequence of the pollution. In addition, there is no complementary provision which deals with the cleaning of the water source. Suriname in this instance does not apply the polluter pays principle. What should also be noted about these two provisions is that they are only concerned with protecting the health and lives of humans. The protection of the environment, including wildlife, is not included. The effects which pollution can have on aquatic life has not been taken into account.

Suriname has regulations on mining-use of mercury and heavy metals. The Mining Decree[1136] concerns the reconnaissance, exploration and exploitation of all mineral resources in Suriname. All raw material in and above the ground is considered property of the state.[1137] When activities take place, the best available techniques should be used and ecosystems should be protected. After exploration has taken place the land should be rehabilitated. The use of mercury is not prohibited by law, however the import of mercury has been regulated.[1138] OGS is the authorised agency to enforce the Mining Decree.

1134 Constitution of the Republic of Suriname (available on: www.constitution.org/cons/suriname.htm last visited: 16 March 2014)

1135 Article 224 Wetboek van Strafrecht G.B. 1911 No. 1 zlg. Bij S.B. 2004 no. 105.

1136 Decreet Mijnbouw S.B. 1986 no. 28, S.B. 1997 no. 44.

1137 Article 2.

1138 Nancy Del Prado, 'An Analysis of Land Rights of the Indigenous Peoples and Maroons in Suriname. Adaptation of Legislation in Suriname' (2006). p. 43.

Suriname also has a Nature Conservation Act[1139] which authorises the President, by means of a decree, to assign land and water belonging to the state as a nature reserve.[1140] An example of a nature reserve is Brownsberg. The criteria for assigning land as a nature reserve are twofold: firstly, that 'the area in question due to its diverse natural and scenic beauty and/or by the presence of important flora, fauna and geological objects requires protection by the government.' Assigning land as a nature reserve prohibits hunting and fishing from taking place on these reserves.

Suriname currently has sixteen protected areas. Seven of these are located in the coastal area, three on the border of the old coastal plains and savanna belt and five reserves and one nature park are located in the Interior. Among these protected areas are Galibi Nature Reserve and the Bigi Pan Multiple Use Management Area. In total the protected areas cover 12% of the national territory. These areas have been assigned for different reasons, such as the protection of important animals and birds. For instance, the Coppename River Estuary Nature Reserve was established because it is the habitat of a famous colony of herons and red ibises. This protected area has also been included in the Ramsar List of Wetlands of International Importance.

As mentioned, the protection of waterways is extremely important for meeting the needs of those living in these areas. Currently, one of the biggest threats to the environment is caused by the activities of both commercial mining companies and small scale gold miners. As mentioned an estimated 75% of artisan or small scale gold miners come from outside of Suriname, specifically Brazil, while only a quarter belong to maroon communities. When tribal communities undertake small-scale gold mining operations near their own communities, thereby contaminating local water resources, this is often an indirect result of unresolved land rights issues and inequalities in benefit-sharing around development of the Interior.[1141] The State in several instances has granted concessions, which overlap with traditional lands, to private companies without consulting tribal and indigenous communities. Due to at times physical confrontations the government and in certain instances private mining companies designated plots of land for exploration by small goldminers However, in this designation practice this is often not done. This results in gold miners undertaking activities in areas near communities or in Nature Reserves.[1142] An example of this can be found in the Brownsberg Nature Reserve.[1143]

The above illustrates that, at present, Suriname does not have a comprehensive framework dealing with water concerns and this results in four major deficiencies and obstacles for sustainable and equitably access to water. Firstly, the water and sanitation

1139 Nature Conservation Act 1954 – GB 1954 No. 26, as amended SB 1992 No. 80.

1140 Article 1.

1141 Interview with Christopher Healy (Culturecom)

1142 Ibid.

1143 Het natuurpark Brownsberg loopt tegen zijn tragisch eind aan. Goudzoekers schrapen steeds meer van de natuurrijke heuvel weg. Minister Steven Relyveld van Ruimtelijke Ordening Grond & Bosbeheer zelf trekt aan de bel. https://www.waterkant.net/suriname/2015/11/13/surinaams-natuurpark-brownsberg-vreet-weg/

sector is not fully and sufficiently regulated. Certain laws have been drafted over the years, but they are still to enter into effect. Legislation relating to surface water which should address, for instance, the discharging of wastewater is also lacking. In addition, responsibilities concerning the water sector are fragmented across institutions and authorities.

There are however opportunities for improvement. And the outdated rules could even prove to be useful if they are enforced.[1144] The regulatory framework currently lacks adequate remedies and is fragmented. However, the legal framework allows for the designation of certain waterways, such as the Suriname River, as protected areas so that harmful activities can be prohibited. The government could choose to allow subsistence fishing so as not to interfere with the livelihood of indigenous and tribal communities. The regulatory framework of Suriname has not been amended to take heed of international agreements that Suriname has signed. Many laws therefore need reviewing.

The presence of a human right to water at the international level currently seems to be the most comprehensive articulation that Suriname can use as an example. Also, if the legal framework is not upgraded and amended, the national level will not be appropriate for individuals to seek redress for infringements. And therefore individuals will be motivated to access international instruments.

7.5.6 BUILDING BLOCK 6: TRADE-OFFS BETWEEN SOCIAL OBJECTIVES

Assessment criteria: Which trade-offs are being made? How are the costs and benefits distributed?

In order to realise the human right to water for vulnerable groups in Suriname different trade-offs need to be made. It is a process of negotiating trade-offs between stakeholders.[1145] Different roles, authorities and parties should be taken into account. As mentioned, Suriname has explicitly stated its aim to provide access to water for its citizens. This is illustrated by the low prices for public water supply services. However, this low pricing is indirectly affecting the sustainability of these services and the expansion of the infrastructure. Decisions about the financial resources of the SWM mostly require political support from Parliament and the Minister of Natural Resources. As was explained, the SWM has to depend on the Government agreeing to its policies. Because of this it can be difficult to introduce new measures which are politically unpopular.[1146]

1144 Statement by Hans Lim A Po (lawyer) during presentation. www.starnieuws.com/index.php/welcome/index/nieuwsitem/29564.
1145 Nyree Stenekes, 'Sustainability and Participation in the Governing of Water Use: The Case of Urban Water Recycling' (University of New South Wales 2006). p. 15.
1146 Interview with Gloria Sandi (SWM).

Also, the main reason for the SWM and the DWV not providing access to water in informal and occupied settlements is that the SWM and the DWV consider them to be inhabited by illegal occupants who do not have a right to be connected. However, this is not reflected in clear policies by these agencies. The water tariffs should not be subjected to as much political influence as they are now. Other options which would support the self-reliability of the SWM should be sought.

In the case of Brokopondo, as mentioned, small-scale gold mining and other extractive industries contribute significantly to Suriname's GDP. And they provide a direct livelihood for many tribal communities and indirectly for the general public. Therefore, natural resources are of great importance for the economic development of Suriname. Financial benefits should be balanced against other goals and concerns. The main concern in relation to mercury pollution, as a result of mining, is that of the health of individuals and the environment. The immediate effects of using polluted water for domestic purposes, like bathing and drinking, have been noted, such as skin rashes.[1147] And, recently, a study has been initiated on the effects of mercury poison on foetal development.[1148] Many pregnant women still eat predatory fish from the rivers as this is their main source of protein. These fish can contain high levels of mercury, however, which affects the development of their unborn children.[1149] The question of whether the financial benefits outweigh the long-term health consequences for future generations in the Interior should be paramount. In addition, the pressure that this situation might entail for the healthcare system in the future has not been looked into. However, it can be assumed that in villages in the Interior a relatively high number of children will be born with birth defects which require healthcare services that those regions currently do not possess. These healthcare costs could dissuade the Government from taking necessary measures.

Another reason for Suriname to protect its surface water from pollution is the possibility of exporting large quantities of water to countries in the Caribbean region which are experiencing water scarcity, like Barbados.[1150] In 2006 the government started investigating innovative ways of using its water resources for further economic diversification. The sale of bulk water was identified as such an opportunity as other Caribbean countries such as Dominica, St. Lucia and St. Kitts and Nevis are experiencing a water crisis.[1151] Through exporting of bulk water, which would be formulated in a Fresh Water Transfer Allocation Agreement, Suriname could assist in providing these countries with water.

An estimated 151 billion cubic meters of freshwater flows through Suriname into the ocean. Suriname has developed a plan to contain water from this flow; more specifically,

---

1147 Interview with Cheryl Sastro (National Women's Movement).

1148 Onderzoek NL – Academisch Ziekenhuis Paramaribo.

1149 Daniel Peplow and Sarah Augustine, 'Community-Led Assessment of Risk from Exposure to Mercury by Native Amerindian Wayana in Southeast Suriname' (2012) 2012 Journal of Environmental and Public Health 1, p. 6.

1150 Del Prado, 'Naar Geintegreerd Waterbeheer in Suriname'. p. 13.

1151 Www.caribjournal.com/2016/06/30/caribbean-water-problem/

it is targeting the Commewijne River for this purpose. Suriname is intending to export 4,5 billion m3 at 1 USD per 3m generating a revenue of USD 4.5 billion and it expects an economic growth of 89 percent.[1152] Not only does Suriname expect to support its own economy but also that of the Caribbean because Suriname expects that the export of water will increase a sustainable water supply for these states at reduced costs. Suriname argues that using its water resources for economic development will motivate Suriname to treat its water resources sustainably. The funds which can be made available by selling these freshwater resources could be invested in the protection of the environment and to finance measures needed to realise the human right to water for all Suriname's residents.[1153]

Suriname has conducted a feasibility study which studied the infrastructure for water collection, purification and transportation.[1154] Nevertheless, there are still several challenges which need to be considered. The study lacks an environmental impact assessment. Containing water before it reaches the ocean will have its effects on the environment. For instance, the coastal plain complex includes mangrove forests which are important breeding, feeding and nursery grounds for, amongst others, fish, sea turtles and birds.[1155] The containment and diversion of vast amounts of freshwater which are needed for the mangroves to thrive will also affect the biodiversity of the coastal plains. Consequently, it will jeopardise the economic benefits that the coast offers to local communities from activities such as (subsistence) fishing and tourism. In addition, even though the feasibility study takes into account the purification of water, the techniques mentioned are limited to sand filtering and the possible addition of chlorine. This is despite the highest mercury levels being found in the targeted Coppename River according to a study conducted in 2012.[1156] For this reason it is questionable whether the water will be suitable for export.

One of the challenges that Suriname faces is anticipating the effects that its current policies and practice will have in the future. The protection of water resources is becoming more important as ground water resources are threatened and surface water will be used in the future also for the needs of those in the urban areas. The practices and measures that governments take are the results of weighing different interests. However, these interests and their weighing are not always made transparent which makes it challenging for citizens to understand the reasoning behind certain decisions. Making the trade-offs transparent would assist in holding the government accountable and would give stakeholders an insight as into which actor make certain decisions.

1152 Http://suriname.conservation.org/projects/fresh-water-transfer-allocation-agreement-suriname/.

1153 Www.telegraaf.nl/buitenland/24692168/__Suriname_kan_verdienen_aan_export_van_water__.html.

1154 Samenvatting Haalbaarheidsstudie http://suriname.conservation.org/files/FTAA/Public_report_FTAA_Suriname.pdf.

1155 Ministry of Labour Technological Development and Environment. p. 20.

1156 Paul E Ouboter and others, 'Mercury Levels in Pristine and Gold Mining Impacted Aquatic Ecosystems of Suriname, South America' (2012) 41 Ambio 873.

7.5.7 BUILDING BLOCK 7: FINANCIAL ARRANGEMENTS

Assessment criteria: Are the financial arrangement for securing water for domestic purposes for vulnerable groups sustainable and equitable?

To support further developments in providing access to water for domestic purposes the SWM has developed what is known as the Masterplan describing the necessary investments in the supply of drinking water for the coastal zone. It has established 95 percent as the desired coverage for the coastal area and the Masterplan sets out the steps which have to be taken.[1157] Groundwater is currently used as the base material for drinking water due to its lower investment and operational costs as surface water requires more treatment.[1158] The SWM has committed itself to providing affordable access to water. Its aim is to ensure that users do not spend more than 5 percent of their minimum income on water.[1159] The UNDP has advised 3 percent as the benchmark.[1160]

The production of a cubic meter of water costs SRD 3. Before 2016 most households paid around SRD 1,70 per m3.[1161] However, since May 2016 SWM has introduced new tariffs. The table 7.7 shows the tariffs for regular households. With the introduction of the new tariffs different categories have been classified: regular households and outside taps, regular households with a swimming pool, commercial connections, and social and public connections. On average households use 20 m3 per month.[1162]

Table 7.7 Tariff structure per 15 May 2016[1163]

Tariff group 20 & 30	Normal household connection and outside/property taps
< 8m3	SRD 10,00 (standing charge) plus SRD 3,00 (renting cost meter)
9 - 15m3	SRD 1,70 per m3
16 - 22m3	SRD 2,25 per m3
23 - 50m3	SRD 4,50 per m3
> 50m3	SRD 4,75 per m3

These tariffs respect the value that water should be affordable for citizens. This categorisation also respects the principle that the user pays while it still allows for small users to have access to water at a price which is lower than the production

1157 N.V. Surinaamse Waterleiding Maatschappij, 'Masterplan for the Coastal Zone' (2002), p. 3. N.V. Surinaamse Waterleiding Maatschappij, 'Suriname Water Supply Master Plan'.
1158 N.V. Surinaamse Waterleiding Maatschappij, 2002 p. 4.
1159 N.V. Surinaamse Waterleiding Maatschappij, 2002 p. 5.
1160 Inga Winkler, 'Policy Guidance Workshop: The Human Right to Water and Sanitation "Translating Theory into Practice"' (2008), p. 8.
1161 Www.de-surinaamse-krant.com/2015/03/waterbronnen-swm-ten-noorden-van.html.
1162 N.V. Surinaamse Waterleiding Maatschappij, 'Informatiebrochure SWM' 1. p. 3 as of April 2012 20 m3 = 20.000 litres. The human right to water ensures between 50 – 100 litres per person per day translating into between 1.500 and 3.000 litres per month. In Suriname.
1163 N.V. Surinaamse Waterleiding Maatschappij, 'Drinkwatertarieven'.

cost. However, what should be kept in mind is that the social tariff is applied per household/connection and does not take into account the number of persons in that household.

However, the revenues that the SWM earns are insufficient to support the services it provides. For several years, the SWM has not received subsidies from the government.[1164] The SWM is unable to independently cover the costs of the drinking water supply and therefore relies on foreign investors and NGOs.[1165] The SWM prefers loans which are not considered to be too much of a risk and not having high inflation rates.

> '*The annual governmental budget does not allocate capital or operation funds to SWM. Thus far, the majority of capital investments are mainly done on project basis, with the support of (international) donors through the government. A wide intervention of variety of equipment provided from a variety of suppliers paid for by different donors also results in more difficult maintenance and less efficient operations.*'[1166]

Interesting in this respect is the obligation for states to ensure that their financial strategies are in line with the realisation of the human right to water. In designing their budgets states should take into account their immediate obligations concerning this right. States should dedicate funds to capacity building and monitoring. As explained, beyond the immediate obligations, states are required to use the maximum available resources. UNDP recommends governments to spend 1 percent of their GDP on water and sanitation.[1167] If the SWM is to reach coverage targets in a sustainable manner, the government should invest the necessary funds. As the government has the final responsibility it is required to support the SWM. It can raise additional funds by measures such as progressive taxation. The government could also choose to prioritise the water sector and anticipate the short and long-term benefits that this sector can produce.[1168] When these benefits reach especially vulnerable groups this will have a positive effect on healthcare and economic development. Merely stating that the government has insufficient funds is not a justification for inaction.

1164 De 10-DE Afdeling van de begroting van uitgaven en ontvangsten voor het dienstjaar 201 betreffende het MINISTERIE VAN NATUURLIJKE HULPBRONNEN p. 51. www.dna.sr/media/141431/SB_2016_no_83_NH.pdf.

1165 While supplying drinking water at a cost-effective and affordable tariff (5% of the minimum income) the SWM was able to finance a maximum of EURO 26 million through its own means until 2010, with the assistance of its 2 sister water companies in the Netherlands (GWA and Vitens). N.V. Surinaamse Waterleiding Maatschappij, 'Masterplan for the Coastal Zone'. p. 5.

1166 49.

1167 United Nations Development Programme, *Human Development Report 2006 Beyond Scarcity: Power, Poverty and the Global Water Crisis* (2006) <www.undp.org/content/dam/undp/library/corporate/HDR/2006 Global HDR/HDR-2006-Beyond scarcity-Power-poverty-and-the-global-water-crisis.pdf>.

1168 Catarina De Albuquerque and Virginia Roaf, 'Financing', *Realising the human rights to water and sanitation: A handbook by the UN Special Rapporteur Catarina De Albuquerque* (2014), p. 7.

Investments in the development of the water sector in the Interior are also necessary. Financial arrangements for establishing new water supply systems in the Interior fall under the responsibility of the Ministry of Regional Development. In 2016, SRD 2,650,000[1169] was designated for projects concerning waste management and drinking water projects.[1170] The Ministry of Natural Resources also designated SRD 2,020,000 for drinking water projects in the Interior.[1171] Considering that its GDP for 2015 was 5.15 billion US dollars the financial investments in the water sector should reach USD 5,150,000 (= SRD 386.652.955).

Besides investments in the water sector the government should provide relevant governmental agencies with the financial means to execute proper monitoring and enforcement. The Gold Sector Regulation Committee (OGS) seems to have the mandate to collect gold tax from miners who have registered under the current small-scale gold mining regulations. However, it is still unclear what its tasks are exactly and where the funds are and will be invested. The Ministry of Natural Resources allocated SRD 5,023,000[1172] to this agency in 2016.[1173]

Suriname has been financially supported by several NGOs in attaining certain goals in the water sector. This mostly takes place on an ad hoc basis and does not enhance sustainability.

Overall funds should be made available for combating environmental degradation and cleaning up pollution. In addition, appropriate filtering systems and rain harvesting systems should be financed for those who cannot afford it themselves. Information about financial arrangements is not made readily available. National budgets are often made available, but after the fact information about how the funds have been used is inaccessible.

Suriname is experiencing a financial crisis[1174] and due to this situation it is advisable that expenses and income should be monitored. The monitoring of state budgets can only be effective if the bodies responsible for this task are independent from state interference.[1175] Monitoring the expenses enhances the accountability of the state and required justification for expenditures. Citizens would be able to monitor where

1169 $352.965 according to.

1170 Www.dna.sr/media/141413/SB_2016_no_77_RO.pdf p. 17.

1171 Www.dna.sr/media/141431/SB_2016_no_83_NH.pdf p. 2.

1172 $ 669.035.

1173 Www.dna.sr/media/141431/SB_2016_no_83_NH.pdf p. 2.

1174 'Suriname is in an economic crisis triggered by a significant commodity terms of trade shock and exacerbated by insufficient buffers and policy responses. The drop in international gold and oil prices and the cessation of alumina production resulted in large fiscal and current account deficits and the onset of a deep recession in 2015. During the boom, there was no institutional arrangement to save resources for future price corrections, and implementation of IMF advice on strengthening the policy framework was limited. Suriname has thus had a much sharper recession, steeper exchange rate depreciation, and larger rise in inflation and government debt than most commodity exporters.' 2016 IMF Article IV report on Suriname available on: www.imf.org/en/News/Articles/2017/01/24/PR1714-Suriname-IMF-Executive-Board-Concludes-2016-Article-IV-Consultation.

1175 Ibid. p. 37.

and how the state's financial resources are spent. This in turn encourages sustainable management of resources.

7.5.8 BUILDING BLOCK 8: ENGINEERING AND MONITORING

Are systems and existing infrastructures being engineered in a sustainable manner? Is there sufficient monitoring of the system and are the data analysed?

Two of the issues undermining access to water for vulnerable groups are related to engineering: water pressure and interruptions to the water supply. The SWM supplies water through pipelines which reach houses or plots of land. However, the pressure with which the water is supplied is low.[1176] This makes it challenging for users to receive enough water to meet their needs. In order to deal with this challenge most water users have installed a hydrophore to pump the water at the desired pressure and private storage tanks which fill up during the night.[1177] The SMW estimates that between 40 to 50 percent of consumers have had storage tanks and hydrophores installed.[1178] These options are only available to those who can afford them. In many of the lower income households and neighbourhoods hydrophores are not present.[1179] It takes these households a relatively long time to receive enough water to meet the water needs of every member of the household.

A second issue is that water users can experience interruptions in their water supply. These interruptions are often connected to power shortages. The SWM depends on hydroelectricity from the Afobaka power plant and reservoir located in the Brokopondo district. The availability of water in this reservoir has been fluctuating during the past few years, especially during times of drought, resulting in the unreliable availability of electricity.[1180] The limited company Energie Bedrijven Suriname (EBS) and the government are looking at possible solutions, for example diverting additional rivers to the reservoir.[1181]

One of the major challenges which the SWM is facing is the amount of Non-Revenue Water (NRW). It is estimated that 40% of water produced by the SWM does not reach the user.[1182] This is caused by faulty infrastructure. However, the pumping stations have been improved and they are currently working properly as table 7.8 illustrates.

1176 Interview with Alwin Linger (SWM)
See also N.V. Surinaamse Waterleiding Maatschappij, 'Suriname Water Supply Master Plan'. p. 13 – 14. 'pressures range from 25m of pressure head at the pumps to as low as 0 – 2m of pressure head at the end point in the system.'

1177 Jack Menke (ed.), *Mozaiek van Het Surinaamse Volk* (2016), p.264.

1178 N.V. Surinaamse Waterleiding Maatschappij, 'Suriname Water Supply Master Plan'. p. 13.

1179 Jurna, p. 51.

1180 Http://nos.nl/artikel/689428-suriname-kampt-met-stroomtekort.html.

1181 See LW Boksteen, 'Strategische Analyse Cn Participatief Actieplan Voor Zuid Oost Suriname' (2009).

1182 Www.de-surinaamse-krant.com/2015/03/waterbronnen-swm-ten-noorden-van.html.
N.V. Surinaamse Waterleiding Maatschappij, 'Suriname Water Supply Master Plan'. p. 49.

Table 7.8 Summary assessment of pumping stations in the Paramaribo District

Summary assessment of pumping stations in the Paramaribo District							
Station	Owner	Power Utility	Electrical installations	Pump installations	Piping	Mechanical	Conclusion and Recommendation
Benny's park (booster)	SWM	Good	Poor	Reasonable	Good	Poor	The booster station needs to refurbish.
Blauwgrond	SWM	Good	Good	Good	Good	Good	SWM is expending the capacity of the station
Flora	SWM	Good	Reasonable	Good	Good	Good	Safety wise the electrical installation needs to refurbished.
Leysweg	SWM	Good	Good	Good	Good	Good	Some part of the electrical installation need to refurbish.
Livomo	SWM	Good	Good	Reasonable	Good	Good	The pumps and motors are outside and are oxidized
Tourtonne	SWM	Good	Good	Good	Good	Good	The station is in good condition.
WK plein	SWM	Good	Good	Good	Reasonable	Reasonable	SWM will replace the aeration unit and some pipes are oxidized and leaking.

Flash floods are a recurring issue in Paramaribo. The coastal zone where Paramaribo is located is flat and during the big rainy season from April to August this region is highly vulnerable to flash floods. These flash floods threaten the general water supply. The Meteorological Service of Suriname (MSS) has determined that there have been peaks in monthly, daily and hourly rainfall data.[1183] The increased intensity of rainfall could be caused by climate change and therefore it can be expected to remain as a recurring issue. The water infrastructure and thereby the water quality are vulnerable when flash floods occur. The SWM should therefore remain vigilant and the government should take water capture into account in its planning. Also, when financial resources are available the SWM should invest in improving water pressure. This will assist poorer households in meeting their water needs.

As mentioned, there are several villages in Brokopondo which still depend on traditional sources of water. When these sources become polluted, more reliable and safe alternatives should be made available. For instance, in the village of Asigron the Government has set up a rain harvesting pilot project.[1184] Another strategy which can be adopted in a situation where mercury has polluted traditional sources is the instalment of equipment to remove mercury.[1185] One of the issues currently being experienced

1183 Meteorological Service Suriname, 'Flash Flood Guidance' 1 <www.wmo.int/pages/prog/hwrp/documents/FFI/Flash_Flood_Guidance_Systems_Background_Information.pdf>.

1184 Nieuwsartikel aftrap rainharvesting project.

1185 *'Conventional chemical coagulation, sedimentation and filtration can achieve removals of up to 80% for inorganic mercury, but only 20–40% for organic mercury. Ferric sulfate is more effective than*

is that the filtering systems which have been installed are not effective in removing mercury. Over half of the population in the Interior have no means of excreta disposal; excreta mostly go straight into the rivers and creeks which at the same time serve as the main water supply. For this reason, appropriate sanitation services should be found in order to protect water resources from human waste. Also, the use of closed systems or recycling mercury is not common practice in Suriname and this could be encouraged if small-scale minors receive appropriate training.[1186] Preventing mercury from entering the waterways in this manner by providing appropriate filtering systems could also be an effective strategy. Suriname has also tried to introduce simple equipment for artisan gold miners which would make the use of mercury unnecessary.[1187]

Meteorological and hydrological observations are of great importance for good water governance. The Meteorological Service of Suriname and the University of Suriname are responsible for monitoring these observations.[1188] However, the data needed for analysis is limited. One of the explanations for this is the Interior War which took place in the 1980s. At the beginning of that war the meteorological network of Suriname included around 183 rainfall stations with 26 climatologic stations and 8 synoptic stations. When the war ended 48 stations remained. With the economic recession that followed and the lack of political will, this number of stations remains to this day and the lack of data collection is aggravated by the inconsistent operation of the stations. Therefore, data collection has been poor as data series are left incomplete.[1189] Other data such as precipitation and river discharge have been consistent over certain periods of time allowing for researchers to analyse and use this information.[1190] This data supports planning the allocation of water resources.

7.5.9 ENFORCEMENT

Assessment criteria: Are regulations and agreements enforceable by public and/or private parties, and are there appropriate remedies available?

aluminium sulfate, and removal is more effective in the presence of high concentrations of suspended solids. Powdered activated carbon is effective for the removal of inorganic and organic mercury and can be used to enhance removal during coagulation. Granular activated carbon treatment is also effective (Sorg, 1979). Ion exchange could be an alternative method (Chiarle et al., 2000). It should, therefore, be possible to achieve a concentration below 1 µg/litre by treatment of raw waters that are not grossly contaminated with mercury.' World Health Organization, 'Mercury in Drinking-Water. Background Document for Development of WHO Guidelines for Drinking-Water Quality' (2005).

1186 Heemskerk and Kooye. p. 11.

1187 Interview with Christopher Healy (Culturecom).

1188 Interview with Sieuwnath Naipal (ADEK).

1189 Riad J Nurmohamed, 'The Impact of Climate Change and Climate Variability on Th Water Resources in Suriname' (Anton de Kom Universiteit van Suriname 2008), p. 24.

1190 See (Nurmohamed, 2008) who used precipitation, temperature and river discharge series in his thesis: the impact of climate change and climate variability on the water resources in Suriname. Other indicators such as evaporation and wind speed were not collected over a longer period of time. They lacked data and were not up to date.

Rules which apply to SWM can have not yet entered into force. However, the Ministry of Natural Resources does have the authority to supervise measures and actions taken by SMW. What could be advised is that during the transition SWM should start implementing the draft legislation or even follow recommendations by UN agencies which set higher standards.

Currently, the Surinamese regulatory framework includes several provisions which could assist indirectly to the protection of water resources in the Interior which have been mentioned in the previous building block on regulations and agreements. The provisions include licenses, punitive and reparative sanctions and other measures. Governmental agencies for instance need the necessary mandates to collect evidence against suspects.

The Nuisance Act includes a licensing system for enterprises[1191] which are assumed to undertake activities which may cause damage to the environment such are bakeries, breweries and wood mills.[1192] Article 17 allows authorised agents to enter spaces for investigative purposes.[1193] This Act allows for the licenses to be revoked when the conditions of the license are not met. The Act however, does not demand the restoration of the situation and the provisions are outdated as they still mention the previous currency of Suriname.[1194] The enforcement of this law would not assist substantially in the protection of water resources.

If enforcement is lacking in water management and governance this may result in inconsistency and conflicts. The enforcement of regulations and provisions in water management and governance is challenging in Suriname. The relevant governmental authorities have been accused of the arbitrary enforcement of legal provisions, for instance when it comes to enforcement in the gold sector by the OGS. For decades, the government was inactive in enforcing any mining regulations pertaining to illegal small-scale mining.[1195] This period of inactivity ended with the establishment of the OGS which was established to bring illegal activities to an end or to bring them within the legal sphere. Those who are confronted with enforcement by the OGS argue that this department only applies the rules against those who do not have political ties with certain political parties. Another issue with OGS is that it misses the needed capacity

1191 Article 6.

1192 Article 2 WET van 27 December 1929 tot vaststelling van bepalingen omtrent het oprichten van inrichtingen, welke gevaar, schade of hinder kunnen veroorzaken (G.B.1930 no. 64), gelijk zij luidt na de daarin aangebracht wijzigingen bij G.B. 1943 no. 119, G.B. 1944 no. 57, G.B. 1944 no. 153, G.B. 1972 no. 96, S.B. 2001 no. 63.

1193 'Zij, die krachtens dit artikel een inrichting binnentreden, zijn, op verzoek van het hoofd of de bestuurder der onderneming, verplicht tot geheimhouding van hetgeen het daarin uitgeoefende bedrijf betreft, voorzover dit niet met de naleving der gestelde voorwaarden in verband staat.'

1194 See for instance article 16: Hij, die opzettelijk de in het slot van artikel 17 opgelegde geheimhouding schendt, wordt gestraft met gevangenisstraf van ten hoogste zes maanden of geldboete van ten hoogste Sf. 500.000,- (vijfhonderdduizend gulden), met of zonder ontzetting van het recht om ambten of bepaalde ambten te bekleden.

1195 Marjo De Theije and Marieke Heemskerk, 'Moving Frontiers in the Amazon: Brazilian Small-Scale Gold Miners in Suriname' [2009] European Review of Latin American and Caribbean Studies/Revista Europea de Estudios Latinoamericanos y del Caribe 5, p. 17.

and means to enforce regulations.[1196] The regulatory system of the gold sector was devised to be self-sustaining as was explained in the building block on means and authority. However, its monetary funds which were supposed to come in have been reported to have gone missing.[1197] In addition, as OGS increases its enforcement illegal mining activities start moving deeper into the Interior.

Because the provisions of administrative law in Suriname are outdated and consistent use of definitions has been lacking, it could be advised to review this section of the law.[1198] A general administrative law act could support the overall development of administrative measures and sanctions and the enforcement of environmental law such as restorative measures and administrative enforcement. It would also provide a general framework on which citizens could depend. Where necessary the Surinamese government could establish special provision relating to specific situations such as environmental regulations.

The entry into force of the Water laws and the Environment Act would provide opportunity for the government to enforce necessary standards. The Environment Act would assist them to determine at an early stage how certain activities will affect the environment and which mitigation measures should be taken.[1199]

Besides OGS it can be advised for an agency to monitor the overall quality of the environment. NIMOS seems to have the expertise but does not have any legal authority or mandate to combat environmental degradation. This could be adjusted once the Environment Act enters into force. One of the gaps which can be noticed and was already mentioned in relation to public participation, is the lack of opportunity for citizens to encourage enforcement by the government. For instance, if the government refuses to use its mandate to prosecute violators, citizens should be able to protest by using a legal remedy.

7.5.10 BUILDING BLOCK 10: CONFLICT PREVENTION AND RESOLUTION

Assessment criteria: Are there sufficient conflict prevention and resolution mechanisms in place?

Paramaribo

The conditions and terms concerning the public water supply explain that in the event of conflicts between users and the SWM, the Minister of Natural Resources will resolve

1196 Ibid. p. 1.
1197 Www.worldnieuws.com/dagblad-suriname/transparantie-goudsector-nog-steeds-zoek/
1198 MR Hoever-Venoaks and LJA Damen, *Surinaams Bestuursrecht* (Ars Aequi Libri 2003). p. 42.
1199 Del Prado, 'Analysis of Instruments, Principles and Legal Norms in Force on Domestic Law of the Amazon Basin in Water Resources Management'. p. 26.

the conflict.[1200] It is unclear, however, how this conflict resolution will take place. In addition, it seems that this course of action has not yet taken place. The SWM has made use of the judicial system in order to resolve its conflicts with consumers.[1201] Information should be made available to consumers in order to clarify which steps they can take to resolve conflicts with the SWM. Also, there should be a means by which those who are currently unserved can communicate with SWM.

Brokopondo

The communities in Brokopondo depend on the government for their access to water as the SWM does not cover this region. Therefore, if a conflict occurs concerning their access, they need to be able to address this through the available channels. This proves challenging, however, especially since Suriname does not have a General Administrative Law Act which regulates communication with governmental agencies.[1202] Nevertheless, there are several avenues available.

Firstly, according to Article 22 of the Constitution, individuals have the right to 'submit written petitions to the competent authority.' However, there is currently no law which explains the appeal or petition procedure.[1203] Laws sometimes include provisions which explain which authority is competent to receive complaints on a specific issue. These authorities are mostly the relevant Ministries and not independent arbitrators.

The government should also have effective measures in place which protect the human right to water for these communities and individuals from infringements by third parties. Individuals could use tort law based on Article 1386 of the Surinamese Civil Code. However, this approach has several obstacles. Another option would be to depend on the Constitutional Court to assess government acts and legislation in relation to constitutional rights.[1204] This Court has not yet been established, however. Nevertheless Article 144 of the Constitution reads as follows:

1. *There shall be a Constitutional Court which is an independent body composed of a President, Vice-President and three members, who – as well as the three deputy members – shall be appointed for a period of five years at the recommendation of the National Assembly.*
2. *The tasks of the Constitutional Court shall be:*
    a. *to verify the purport of Acts or parts thereof against the Constitution, and against applicable agreements concluded with other states and with international organisations;*

1200 N.V. Surinaamse Waterleiding Maatschappij, 'Voorwaarden Voor Aansluiting En Waterlevering' 1. article 10.
1201 Interview with Gloria Sandie (SWM).
1202 Hoever-Venoaks and Damen.
1203 Ibid. p. 42.
1204 Article 144 Constitution and see www.dna.sr/media/64070/Wet_Constitutioneel_Hof.pdf.

b. to assess the consistency of decisions of government institutions with one or more of the constitutional rights mentioned in Chapter V.

3. In case the Constitutional Court decides that a contradiction exists with one or more provisions of the Constitution or an agreement as referred to in paragraph 2 sub a, the Act or parts thereof, or those decisions of the government institutions shall not be considered binding.

4. Further rules and regulations concerning the composition, the organisation and procedures of the Court, as well as the legal consequences of the decisions of the Constitutional Court, shall be determined by law.

This Court would have the authority to declare legislation non-binding if it finds that it infringes constitutional rights. However, this would not result in new legislation. And the Court would have to interpret the right to water as part of a constitutional right such as the right to life. The establishment of a Constitutional Court could greatly contribute to the protection of human rights in Suriname and would respect the right of access to justice for individuals and groups. This subject has become highly politicised during the past few years.[1205] Nevertheless, an independent Constitutional Court could assist vulnerable groups in protecting their human rights and balance their interests against that of the majority.[1206]

If individuals can eventually find their way to an independent judge, then having their case dealt with in a timely fashion can be challenging as the Surinamese judicial system often has long waiting times due to the courts' lack of capacity.[1207]

The tribal legal system is often used to solve local disagreements. However, it has not been effective in settling conflicts between gold miners, the government and large extractive companies.[1208] However, it would be a possibility for the tribal legal system to encourage one of the players, namely small-scale gold miners, to use sustainable methods such as mercury recycling and closed mining systems.

Suriname should assess its public administrative law provisions and consider implementing a general administrative law act which includes sanctions for environmental degradation but also shaped how citizens can participate and conflicts are resolved as was explained in the previous building blocks.[1209] With the adoption of new legislation, older legislation may be complementary and should be reassessed in order to ensure disparities in protection do no occur. For complainants, it should be relatively easy to discern where they can file their complaint and how.

1205 Www.volkskrant.nl/buitenland/amnestiewet-toetsen-door-constitutioneel-hof-dat-alleen-nog-niet-bestaat~a3240356/
www.dbsuriname.com/dbsuriname/index.php/vrees-voor-politisering-constitutioneel-hof/

1206 Kambel and MacKay, p. 134.

1207 Https://www.waterkant.net/suriname/2016/08/25/snel-nieuwe-surinaamse-rechters-korte-cursus/

1208 Heemskerk and Kooye, p. 12.

1209 Hoever-Venoaks and Damen.

7.6 RECOMMENDATIONS

Table 7.9 Overview strengths and weaknesses exposed by assessment framework

	Paramaribo	Brokopondo
Knowledge		
Values/principles		
Stakeholders		
Trade-offs		
Authority & means		
Regulations		
Finances		
Engineering		
Enforcement & Monitoring		
Conflict remedies		

Table 7.9 illustrates the strengths and weaknesses in the overall water system, following the assessment in this chapter of sustainable access to water for vulnerable groups in Suriname. Green indicates strength, orange indicates the presence of weaknesses while there is movement towards strength and lastly red indicates overwhelming weaknesses and an immediate need for improvement. This table clearly shows that there is room for improvement. The knowledge which SWM possesses can be seen as contributing to the realisation of a sustainable human right to water. SWM has access to information about Suriname's groundwater resources and seems to monitor developments. In the Interior, on the other hand, most of the conclusions concerning water quality are the result of outdated information and observational conclusions are made, for instance, in relation to communities using water ways for sanitation purposes. Not one specific organisation is responsible for continuously monitoring water quality in the Interior. The government, in cooperation with the University of Suriname as it has the knowledge and capacity, should coordinate the monitoring of water quality in Suriname in general. It seems, however, that this should be supported by capacity-building and access to technology. As was explained during fieldwork in Nickerie where the water quality is sometimes tested, the samples are send to laboratories in the Netherlands and often the financial resources are often lacking. Maybe cooperation at the local level within the Caribbean region could also provide a solution. Other states in the region might have the expertise and equipment which are necessary.

In relation to all other building blocks, improvements are necessary and some, such as the regulations like the water laws, are already on the right track. The adoption of the draft legislation can be a step in the right direction. The drafting process has already taken several years and now it is necessary to adopt the legislation, even though it is not perfect and still has gaps. Adopting the legislation will provide a blank canvas on which to paint on and adjust legislation and policy as necessary. It can even be the case so that after the entry into force of the laws and their implementation and enforcement, the understanding of the situation in Suriname will be enhanced and will feed back into policy, legislation and engineering by SWM. The company is aware of where faulty infrastructure should be repaired or replaced.

When it comes to the values and principles it is positive to see that everyone agrees that access to water should be available to everyone for domestic purposes. However, how this value is weighed against other values such as economic development and sustainability is challenging. As has been shown, the low tariffs prevent SWM from providing sustainable services as they are not supported by the government. Yet, it is the government, who by using the tariffs as a political tool, which prevents these tariffs from being raised. Of course, this does protect the affordability for vulnerable groups but this can also be done by directly providing subsidies to those groups directly.

The values and principles are an important part of the trade-offs. As Suriname is experiencing austerity, economic growth is high on its agenda. As a result, it tolerates infringements of the right to health of especially tribal and indigenous communities, who already have limited access to health care. The government should not only predict the revenues to be earned from the extractive industries but should also estimate the costs of environmental, social and health side-effects. In this way, the trade-offs can be better substantiated and become transparent. This would provide the government with a clearer view of what is at stake.

The lack of stakeholder participation and the participation of civil society in general is a major issue in Suriname. The case studies show that relevant parties are often treated as merely consumers or they are ignored if they belong to certain vulnerable groups and represent conflicting interests. The government should adopt the measures which it was instructed to adopt by the Inter-American Court in relation to tribal and indigenous people. In addition, the avenues that citizens can now take do not lead to meaningful participation. When stakeholders participate in, for instance, meetings of the resort councils they do not know how their voice influences the outcome and what weight their voice has. This can be altered by enhancing the transparency of decision-making at the central level. In addition, the voices of the community should be given a clear weight.

Suriname could choose to establish Water Boards where stakeholders can have a direct impact on water management. This would also allow all relevant stakeholders to participate and to resolve conflicts of interests out. In addition, if the Water Boards would oversee operations at the basin level, stakeholders would have to cooperate and deal with the challenges, such as mercury pollution at its core.

What also works in the context of Suriname is the approach by which access to water in the coastal areas is the task of one agency while access to water in the Interior is the task of another, is a good approach. Because of this both organisations (DWV and SWM) can focus on the special needs of their particular area. This allows them to deepen their knowledge and expand their capacity.

Suriname is not meeting its financial obligations in supporting the right to water. SWM seems to have been left in isolation and is struggling to make ends meet. As a result, the sustainability of services is threatened. Suriname should aim to invest at least 1 percent of its GDP in towards realising this right and this should be done in a transparent manner. This transparency should be followed up in the monitoring as well. An independent institution should review the income and expenses of the government, so that the government can be held accountable and that resources are spent equitably.

7.7 CONCLUSION

The case studies of two districts in Suriname enhance the understanding of what is needed for the human right to water to be realised in a sustainable manner. Localising human rights brings along many challenges which might not have been noticed while formulating the right at the international level. The case studies demonstrate that the realisation of the human right to water may require different means depending on the setting. In this regard, the environment in which individuals live is highly relevant.

This realisation by different measures and strategies can be relevant not only between countries but also within a state and between regions. Brokopondo and Paramaribo might be in the same country and share the same legal framework but the environment plays such a significant role in how the right can be realised as the necessary infrastructure is lacking. The case studies show that there are structural issues within the Surinamese water sector which should be addressed. Improving only one building block will not have the desired effect if the other building blocks are ignored. These building blocks spill over into one another and are interdependent.

CHAPTER 8
CONCLUSIONS

'Water is life. It is the lifeblood of our environment and our very existence.'[1210]

8.1 INTRODUCTION

This study commenced by explaining the situation of the current water crisis where one out of ten persons globally does not have access to safe water for domestic purposes. When this issue is not tackled it will negatively affect generations to come and destroy opportunities for the development of current generations. This deprivation of water is set against a background of water pollution, over-exploitation and mismanagement but also climate change and economic austerity. The current water usage and allocation is not socially or environmentally sustainable. What should be kept in mind is that there is enough freshwater available to meet the basic needs of current and future populations. However, insufficient water is allocated for domestic purposes. These circumstances exacerbate the challenges that unserved and underserved communities and individuals face. This makes them vulnerable. They make use of unsafe water resources to meet their needs with all the associated health risks.

Human rights aim to provide means of redress, especially for those whose basic needs have not been fulfilled by setting minimum standards. As access to water is the latest right which has been implicitly recognised, its placement within the human rights framework gives vulnerable groups another means to gain access to water in a meaningful way. However, because this right is not explicitly included in the ICCPR and ICESCR its articulation is still at the development stage. As mentioned, access to water is challenging because of these wide-ranging water problems. Therefore for the right to be meaningfully realised appropriate measures should be taken. Such appropriate measures can be found in the concept of sustainable development as it governs how water resources can be managed and allocated. Sustainable development can support the realisation of the human right to water and therefore this study has delved into the main research question: how can the concept of sustainable development contribute to the sustainable realisation of the human right to water for vulnerable people?

1210 Archer. p.2.

8.2 WIDENING THE SCOPE OF VULNERABILITY

The first step which this study took was to look into 'vulnerability' as those who are underserved and unserved can be qualified as vulnerable. Not only are they the most relevant group in the perspective of this study, their situation also serves as a litmus test for the overall realisation of the human right to water. If their access has been guaranteed, it can be concluded that all others will most probably have access to water as well since they do not face the obstacles which vulnerable groups have to face. It is important for states to implement necessary measures at this moment in time because it can be expected that the amount of vulnerable people who do not have access to water will most likely increase during the forthcoming years and decades. More individuals and groups will fall below the threshold of a minimum standard of living regarding access to water because of, amongst other things climate change, if interventions do not take place.

The human rights framework should provide an answer to major problems such as climate change and environmental degradation and it should anticipate future issues if it wants to remain relevant and provide protection for vulnerable groups. The human right to water can already be seen as a step forward since it reflects a developing issue which is related to the environment. The human rights framework should also take into account environmental factors which influence the realisation of human rights and this should also be reflected in the application of special measures for vulnerable groups. However, this study shows that the UN ESCR framework, in which the right to water has been placed, does not take into account environmental factors as a cause of or a contributing factor to the vulnerability of individuals and groups. It mainly looks at social, political and economic factors. This approach will be inadequate for protecting 'environmental human rights' such as the right to health and the human right to water.

The analysis in chapter 2 demonstrated that vulnerability is often related to the grounds of discrimination prohibited by the CESCR, such as race and sex. As a result, the ESCR framework adopts a group-specific approach which could become rigid and risks marginalising certain groups, but it also overlooks individuals who could benefit from targeted measures. Environmental degradation highlights that those who are vulnerable are not in this situation merely because they belong to a specific group and therefore have limited access to coping mechanisms, but that they are vulnerable due to the situation that they are in. This study has therefore found that the ESCR could benefit from adopting an intersectional approach. In the case of the human right to water this would mean that not only social, economic and political factors are taken into account when assessing vulnerability but also environmental factors. The CESCR could apply this approach in assessing individual and group complaints as the Optional Protocol of the ESCR has entered into force. This approach would be more appropriate to confront the major problem of environmental degradation. Widening the scope of vulnerability under the ICESCR will assist the overall goal of realising a sustainable human right for vulnerable people. Chapter 2 therefore shows that in order to provide protection for those who are currently unserved and underserved, the human rights framework should take environmental factors into account when assessing the vulnerability of individuals and groups.

8.3 DEVELOPING THE HUMAN RIGHT TO WATER WITHIN THE CONTEXT OF DETERIORATING WATER RESOURCES

The human right to water is a special right, as it is not explicitly mentioned in the catalogue of international human rights, the Universal Declaration of Human Rights and the two corresponding conventions on economic, social and cultural rights and civil and political rights. During the 1990s the increase in the privatisation of public services motivated certain communities to protest against the privatisation of the public water supply services as a means of protecting access to water. In addition, states began to formally recognise the human right to water and the right to a clean and healthy environment in national legislation. These developments were echoed at the international level when the UN General Assembly (politically) recognised the human right to water in 2010 which was followed up by similar actions by the UN Human Rights Council.

However, before access to water was recognised as a human right it was part of the body of international environmental policies. For instance, the 1977 Mara del Plata Declaration states: 'all peoples whatever their stage of development and their social and economic conditions, have the right to have access to drinking water in quantities and of a quality equal to their basic needs.'[1211] This right made its way to the human right framework but without UN resolutions explicitly including environmental aspects. However, placing this right in an environmental context is crucial for its realisation.

Because the human right to water is not explicitly recognised, in order to clarify its content and states' obligations the UN Committee on Economic, Social and Cultural Rights took the initiative to place the human right to water within the scope of the ICESCR. This does not preclude that the right can also be placed within the scope of the ICCPR as falling under the right to life and the right to be free from cruel, inhuman and degrading treatment. The authoritative but not legally binding interpretation of the right provided by the CESCR in General Comment no. 15 states:

> '*The human right to water entitles everyone to sufficient, safe, acceptable, physically accessible and affordable water for personal and domestic uses. An adequate amount of safe water is necessary to prevent death from dehydration, to reduce the risk of water-related disease and to provide for consumption, cooking, personal and domestic hygienic requirements*'[1212]

1211 Report of the United Nations Water Conference, Mar del Plata, 14–25 March 1977 (United Nations publication, Sales Np. E.77.II.A.12), part 1, chapter I, section C, para. 35.

1212 Committee on Economic, Social and Cultural Rights, 'General Comment No. 15 The Right to Water' (2002). par. 2.

The right is self-standing and it is also an essential element of other rights such as the right to an adequate standard of living and relates to the right to the highest attainable standard of health and the rights to adequate housing and adequate food.

The normative content of the right clarifies which uses are covered by this right. For example, agricultural uses are excluded, even though some argue that water for farming should be included. They do so because it would be more in line with the lived experience of underserved people. The normative content, as developed by the UNCESCR, also explains that water should be accessible in all aspects of daily living and that physical safety should be guaranteed especially for women and girls. General Comment no. 15 mentions the normative content and directs states towards standards provided by authoritative organisations such as the World Health Organization for specifying, for instance, the quality standards that drinking water should meet.

The interpretation provided by the CESCR greatly contributes to the development of the human right to water. As a human right, a range of state obligations accompany this right. States are required to respect, protect and fulfil the right. States should also intervene when third parties infringe on the rights of individuals and oversee the realisation of the right if private companies play a part therein. States are also obliged to realise this right progressively and without discrimination. The CESCR has applied its core content approach to the human right to water also by explaining that the right has minimum requirements that should be met at all times for all individuals.

The implementation of the right at the national level has also contributed to the development of the human right to water at the international level. The national perspective shows that the articulation of the human right to water confronts the domestic challenges that states experience in relation to water, such as droughts and pollution. For instance, the formulation of India's human right to water reflects its problem with the large-scale pollution of water resources. And states like South Africa and Mozambique have chosen to prioritise water allocation for domestic purposes. This highlights the fact that for the human right to water to be meaningfully realised, the management of water resources is highly relevant. One of the major contributions of the human right to water is that it can influence water management by supporting the prioritisation of the water that is available for personal and domestic uses and protects water resources from pollution and over-exploitation. This supports local, unserved communities in safely accessing water locally.

8.4 SUSTAINABLE DEVELOPMENT

Realising the human right to water for vulnerable groups requires a sustainable approach as water resources should be managed in a sustainable manner in order to support the human right to water. The CESCR refers to sustainable development and sustainability in relation to the human right to water on several occasions in General Comment no. 15. However, it does not elaborate on what sustainability actually means for the human right to water. Chapter 4 analysed what this elusive concept entails. The most authoritative interpretation of sustainable development is to be found in the Brundtland Report and it explains sustainable development as follows: 'development that meets the needs of the present without compromising the ability of future generations to meet their own needs.'[1213]

The operationalisation of sustainable development is assisted by different principles which have been developed. There are 7 main principles which provide a further articulation of sustainable development and can be applied to water resources: the duty of States to ensure the sustainable use of natural resources; equity and the eradication of poverty; common but differentiated obligations, the precautionary approach to human health, natural resources, and ecosystems; public participation and access to information and justice; good governance; integration and interrelationship. These principles tell us how the human right to water can be interpreted in such a manner that it will be realised in a sustainable way.

Sustainable development allows for the weighing of different interests and takes into account environmental, social and economic factors. However, in practice, measures under sustainable development are often motivated by economic growth. Sustainable development as a concept is in itself also still at the development stage and is shaped by both soft and hard law instruments such as the Rio Declaration and the ILA New Delhi Principles. These instruments do no create options for redress for individuals as they are soft law in nature.

The sustainable management of water resources allows differences in needs between groundwater and surface water to be recognised. These two types of resources require different kinds of measures and are regulated by different regimes. Analysing these regulatory systems in relation to the human right to water will assist in a better alignment of the use of water resources.

8.5 CONSTRUCTING A SUSTAINABLE HUMAN RIGHT TO WATER

Sustainable development can be complementary and supportive of human rights in general and the human right to water in particular. The obligations which stem from sustainable development can be seen as contributing factors which respect, protect and fulfil the right to water. Both frameworks require access to information and participation by stakeholders. In addition, they aim for equity and seek to facilitate human development through governance measures. Besides these complementary factors there are also tensions which have to be reconciled and they can expand the human right to water and thereby enhance it. The human right to water is focused on the individual while sustainable development focusses on the collective which allows

1213 Brundtland Commission.

for a balancing of interests on a larger scale. Human rights are moving towards a more collective perspective where appropriate. It allows for group complaints, recognises group rights and takes group-specific measures as explained before. The collective approach will aid environmental rights.

Chapter 4 demonstrated that the enforcement of sustainable development is challenging. Principles and requirements are not accessible to individuals other than the rights to participate, and to have access to information and justice. Using elements and principles of sustainable development to enhance the human right to water will allow individuals to gain access to them and to depend on them. They may even gain a legally binding status as part of, for instance, customary law when national and international judicial bodies use sustainable development as part of their reasoning as has been done before by the ICJ.[1214]

Sustainable development is a concept which can be applied across water uses. The human right to water focusses on water for domestic purposes, however. Sustainable development in relation to water resources provides space for prioritising water for basic needs. In conjunction with this the human rights framework requires states to allocate water in such a manner that the survival level of the different rights which require access to water is reached. States should take into account the levels of realisation in allocating water. For instance, if water is allocated to growing non-essential crops in the agricultural sector while communities are at risk of dying due to dehydration, then this water should be reallocated.

Obligations regarding the protection and management of water resources can be seen as remote requirements which protect and fulfil the human right to water. These especially affect the following elements: international cooperation and common but differentiated responsibilities, the priority of use, reallocation and allocation measures. The substantive elements of the human right to water, especially the requirement of using *'all available resources'* for the realisation of human rights, can be further developed by considering natural resources such as water resources. Instead of requiring states to use all available natural resources, the CESCR should use the principle of sustainable use to assess whether states are meeting their obligations.

8.6 CROSS-CUTTING THEME NEEDS A MULTIDISCIPLINARY APPROACH

In order to analyse the challenges and opportunities for the realisation of a sustainable human right to water for vulnerable groups at the national level it was necessary to formulate a multidisciplinary framework which systematically studied the aspects which influence the management of water resources and how water services reach

1214 *Case Concerning Military and Paramilitary Activities In and Against Nicaragua (Nicaragua v. United States of America); Merits*, International Court of Justice (ICJ), 27 June 1986.

vulnerable groups. The 'Ten Building Blocks framework'[1215], as developed by van Rijswick, Edelenbos, Hellegers, Kok, & Kuks, was adjusted to take into account the core aspects of a sustainable human right to water as was discussed in chapters 3 and 5. This multi-disciplinary approach provides insights into aspects outside of the legal field which can assist in substantively realising the human right to water for vulnerable people in a sustainable manner.

8.7 SURINAME

The case study clearly shows that the management of water resources (groundwater and surface water) is directly connected to the sustainable realisation of the human right to water. Even though the public water supply in Suriname depends on groundwater, surface water will become more relevant for public water supply in the future and it should therefore remain clean. Even now surface water is the main source of water for domestic purposes for those living in the Interior. The challenges faced by the inhabitants of the Interior regarding access to water reflects their vulnerability as they are highly dependent on the natural environment. Even though the majority of those residing in the Interior belong to tribal and indigenous communities and can be considered vulnerable, the case studies makes clear that when it comes to environmental degradation group-specific approaches will only have limited effect. Vulnerable people should be able to facilitate their own access in the Interior if proper filtering systems and rain harvesting systems will be provided. Individuals should be given different options in the Interior to meet their needs. For instance, different levels of water quality can be used for different uses. Also, the implemented systems should adapt to seasonal changes and climate change which influence the availability of fresh water.

The case study also makes clear the importance of governance. The regulatory framework in Suriname is outdated, fragmented and incomplete. However, in certain instances even though mandates might be lacking, certain organisations do step in where necessary. This is for instance the case in the water supply by the SWM and its water quality control. Overall the enforcement of the current legislation which could be effectively used is not taking place. And due to the gaps in the regulatory framework and enforcement, water management has become highly politicised. As a result, realising the human right to water does not only depend on the amount of water a state has but also on the weighing of interests which are not always clear and transparent or even legitimate.

The case studies clearly demonstrate that the realisation of the human right to water requires interventions on multiple levels. Adjusting the legal framework is not sufficient if the overall environment is not supportive of a right to water and water resources are not managed sustainably. The case studies also show that localising the human right to

---

1215 Van Rijswick and others.

water requires different measures and strategies in different contexts which happen to be in the same national context. Therefore, applying a nationwide strategy might not be effective or sustainable when states want to meet their obligations.

8.8 FINAL CONCLUSIONS

This study sought to answer a key question: how can the concept of sustainable development contribute to the sustainable realisation of the human right to water for vulnerable people, by taking the following approaches. This was answered in the following ways. Firstly, by widening the scope of vulnerability to include environmental factors. Those who do not have access to water, as guaranteed by the human right to water, will be eligible for targeted measures when environmental vulnerability is considered in conjunction with social vulnerability. Secondly, the principles which accompany the concept of sustainable development can be used to inform and further develop the human right to water. In this way the right will be sustainably realised and states will be able to meet both requirements under international human rights law as well as under environmental law.

What does this mean for the human right to water at the international level? The human right to water was authoritatively interpreted in 2002 and politically accepted in 2010. The core requirements of the human right to water have been established and now attention should be given to aligning other provisions within international law relating to water resources. At the national level a sustainable human right to water will allow for context specifications to be taken into account. Those states which are confronted by, for instance, depletion will be able to justify allocation measures by referring to the human right to water. Also, individuals will be able to hold the state accountable for pollution on account of this right. When implementing the right at the national level, states should realise that it often requires harmonisation and a better coordination of regulations and practices regarding the whole water sector.

With the adoption of the Sustainable Development Goals, universal access has rightfully been placed within the context of water management. These goals can assist in the realisation of the human right to water and can motivate states to adopt the required legislative, technical and financial measures. The SDGs prove that environmental policy and the human rights framework are communicating back and forth on the issue of universal access.

The former Special Rapporteur on the human right to safe drinking water and sanitation (the former Independent Expert) had already set out a course for sustainability as a cross-cutting requirement by establishing it as a fundamental principle for the realisation of human rights.[1216] She has emphasised that sustainability

[1216] De Albuquerque, 'Report of the Special Rapporteur on the Human Right to Safe Drinking Water and Sanitation'. para. 66; In her analysis she used the notion of sustainability as presented in the outcome document of the United Nations Conference on Sustainable Development adopted

means that 'water quality and availability have to be ensured in a way that respects and supports the larger environment. Water contamination and over-extraction must be avoided in order to ensure continued access to safe and sufficient water.'[1217] This course should be continued. For the human right to water this would mean that the elements of this right would be read in tandem and interpreted coherently with the principles of international environmental law.

in 2012 which reaffirms that: 'changing unsustainable and promoting sustainable patterns of consumption and production and protecting and managing the natural resource base of economic and social development are the overarching objectives of and essential requirements for sustainable development'. UN General Assembly (2012) p. 1.

1217 Ibid. para. 66.

BIBLIOGRAPHY

Adebowale M and others, 'Environment and Human Rights: A New Approach to Sustainable Development' [2001] Opinion World Summit on Sustainable Development 1 <http://pubs.iied.org/pdfs/11016IIED.pdf>

Adger WN, 'Vulnerability' (2006) 16 Global Environmental Change 268

Alam S, Klein N and Overland J, *Globalisation and the Quest for Social and Environmental Justice* (2010)

Albuquerque C De, 'Report of the Independent Expert on the Issue of Human Rights Obligations Related to Access to Safe Drinking Water and Sanitation' (2010)

Algemeen Bureau voor de Statistiek – Suriname, 'Resultaten Achtste Volks- En Woningtelling in Suriname (Volume III) Huishoudens, Woonverblijven En Gezinnen, Milieu, Criminaliteit' (2014)

Algemeen Bureau voor de Statistiek – Suriname, Conservation International and UNDP, 'Environment Statistics' (2014)

Alston P and Quinn G, 'The Nature and Scope of States Parties' Obligations under the International Covenant on Economic, Social and Cultural Rights' (1987) 9 Human Rights Quarterly 156

Anand PB, 'Right to Water and Access to Water: An Assessment' (2007) 19 Journal of International Development 511

Anderson AJ and others, 'Integrated Monitoring of Water Allocation Reform in South Africa' (2008) 34 Water SA 731

Anderson W, *Principles of Caribbean Environmental Law* (Environmental Law Institute 2012)

Anonymous, 'What Price for the Priceless? Implementing the Justiciability of the Right to Water' (2007) 120 Harvard Law Review 1067

Antonelli M and Greco F, *The Water We Eat* (Marta Antonelli and Francesca Greco eds, Springer 2015)

Archer JL, 'Rivers, Rights & Reconciliation in British Colombia: Lessons Learned from New Zealand's Whanganui River Agreement' (2014)

Arts K, 'Inclusive Sustainable Development: A Human Rights Perspective' (2017) 24 Current Opinion in Environmental Sustainability 58 <http://linkinghub.elsevier.com/retrieve/pii/S1877343517300301>

Avila-garcía P, 'The Human Face of Water Security' <http://link.springer.com/10.1007/978-3-319-50161-1>

Baba A and others (eds), *Climate Change and Its Effects on Water Resources* (Springer Science Business Media BV 2011)

Backer LC and others, 'Naturally Occurring Water Pollutants' in Janine MH Selendy (ed), *Water and Sanitation-Related Diseases and the Environment: Challenges, Interventions, and Preventive Measures,* (John Wiley & Sons, Inc 2011)

Barnett J, 'Human Rights and Vulnerability to Climate Change' in Stephen Humphreys (ed), *Human Rights and Climate Change* (Cambridge University Press 2010)

Bates R, 'The Road to the Well : An Evaluation of the Customary Right to Water' (2010) 19 Review of European Community and International Environmental Law 282

Benidickson J and others, *Environmental Law and Sustainability after Rio* (Jamie Benidickson and others eds, Edward Elgar Publishing 2011)

Benvenisti E, 'Collective Action in the Utilization of Shared Freshwater: The Challenges of International Water Resources Law' (1996) 90 The American Journal of International Law 384

Bernal C, 'The Right to Water: Constitutional Perspectives from the Global South', *International Environmental Law and the Global South* (2015)

Beyer MG, 'The Global Consultation on Safe Water and Sanitation for the 1990s' (1991) 15 Natural Resources Forum 118

Birnie P, Boyle A and Redgwell C, *International Law and the Environment* (Third, Oxford University Press 2009)

Biswas AK, 'Water as a Human Right in the MENA Region: Challenges and Opportunities' (2007) 23 International Journal of Water Resources Development 209 <www.tandfonline.com/doi/abs/10.1080/07900620701523277> accessed 8 March 2013

Biswas AK, 'Integrated Water Resources Management: Is It Working?' (2008) 24 International Journal of Water Resources Development 5

Björklund G and Sjödin J, *The Human Right to Water and Sanitation Securing Access to Water for Basic Needs* (2010)

Bluemel EB, 'The Implications of Formulating a Human Right to Water' (2004) 31 Ecology Law Quarterly 957 <http://heinonlinebackup.com/hol-cgi-bin/get_pdf.cgi?handle=hein.journals/eclawq31&section=42>

Boksteen LW, 'Strategische Analyse Cn Participatief Actieplan Voor Zuid Oost Suriname' (2009)

Bond JE, 'International Intersectionality: A Theoretical and Pragmatic Exploration of Women's International Human Rights Violations' (2003) 52 Emory Law Journal 71

Bortscheller MJ, 'Equitable But Ineffective: How The Principle Of Common But Differentiated Responsibilities Hobbles The Global Fight Against Climate Change' (2010) 10 Sustainable Development Law & Policy 49 <http://digitalcommons.wcl.american.edu/cgi/viewcontent.cgi?article=1036&context=sdlp>

Bosselmann K, *The Principles of Sustainability* (Ashgate Publishing Limited 2008)

Boucek C, 'Yemen: Avoiding a Downward Spiral' (2009) 102

Bourquain K, *Freshwater Access from a Human Rights Perspective* (Brill Academic Publishers 2008)

Boyd DR, 'Constitutions, Human Rights and the Environment: The Context', *The Environmental Rights Revolution* (UBC Press 2012)

Brklacich M, Chazan M and Bohle H-G, 'Human Security, Vulnerability, and Global Environmental Change' in Richard A Matthew and others (eds), *Global Environmental Change and Human Security* (The MIT Press 2010)

Brooks DB, 'Human Rights to Water in North Africa and the Middle East: What Is New and What Is Not; What Is Important and What Is Not' (2007) 23 International Journal of Water Resources Development 227 <www.informaworld.com/openurl?genre=article&doi=10.1080/07900620601097075&magic=crossref> accessed 8 March 2013

Bruch CE, *Public Participation in the Governance of International Freshwater Resources* (United Nations University Press 2005)

Bruin M, 'Erken Ons VIDS En OIS, de Stem van Inheemsen in Suriname' (Utrecht University 2008)

Brundtland Commission, 'Our Common Future: Report of the World Commission on Environment and Development' (1987)

Buijze AWGJ, 'Thee Principle of Transparency in EU Law' (Utrecht University 2013) <http://dspace.library.uu.nl/handle/1874/269787>

Bulto TS, 'The Emergence of the Human Right to Water in International Human Rights Law : Invention or Discovery?' (2011) 7

Bulto TS, *The Extraterritorial Application of the Human Right to Water in Africa* (Cambridge University Press 2014)

Bungenberg M and Hobe S, 'Permanent Sovereignty over Natural Resources' [2015] Permanent Sovereignty Over Natural Resources 1

Cahill A, '"The Human Right to Water – a Right of Unique Status": The Legal Status and Normative Content of the Right to Water' (2005) 9 The International Journal of Human Rights 389 <www.tandfonline.com/doi/abs/10.1080/13642980500170840> accessed 26 December 2013

Castellino J, 'The MDGs and International Human Rights Law: A View from the Perspective of Minorities and Vulnerable Groups' (2009) 13 The International Journal of Human Rights 10 <www.tandfonline.com/doi/abs/10.1080/13642980802532275> accessed 16 December 2013

Catarina de Albuquerque, 'Chronicle of an Announced Birth: The Coming into Life of the Optional Protocol to the International Covenant on Economic, Social and Cultural Rights – The Missing Piece of the International Bill of Human Rights' (2010) 32 Human Rights Quarterly 144

Cavallo GA, 'The Human Right to Water and Sanitation : From Political Commitments to Customary Rule ?' (2012) 3 Pace International Law Review Online Companion 136

CEDAW Committee, 'General Recommendation No. 25 on Article 4, Paragraph 1, of the Convention on the Elimination of All Forms of Discrimination against Women, on Temporary Special Measures' 10

Center for Strategic and International Studies, 'Global Water Futures. Addressing Our Global Water Future' (2005)

CESCR, 'Guidelines on Treaty-Specific Documents to Be Submitted by States Paries under Articles 16 and 17 of the International Covenant on Economic, Social and Cultural Rights' 15

Chapman AR and Carbonetti B, 'Human Rights Protections for Vulnerable and Disadvantaged Groups: The Contributions of the UN Committee on Economic, Social and Cultural Rights' (2011) 33 Human Rights Quarterly 682 <http://muse.jhu.edu/content/crossref/journals/human_rights_quarterly/v033/33.3.chapman.html> accessed 17 December 2013

Chowdhury N and others, 'The Human Right to Water and the Responsibilities of Businesses: An Analysis of Legal Issues.' (2011) 3

Chuffart S and Vinuales JE, 'From the Other Shore: Economic, Social and Cultural Rights from an International Environmental Law Perspective' in Eibe Reidel and Christophe Golay (eds), *Economic, Social and Cultural Rights in International Law* (Oxford University Press 2014)

Coenen F, 'Local Agenda 21 : " Meaningful and Effective " Participation ?' in Frans Coenen (ed), *Public Participation and Better Environmental Decisions* (Springer Netherlands 2009)

Collin RW and Collin RM, 'Sustainable Development Environmental Justice and Sustainability' in Michael Redclift (ed), *Routledge International Handbook of Sustainable Development* (Taylor and Francis 2015)

Committee on Economic and Social and Cultural Rights, 'CESCR General Comment No. 3: The Nature of States Parties' Obligations (Art. 2, Para. 1, of the Covenant)', vol 1990 (1990)

Committee on Economic Social and Cultural Rights, 'General Comment No. 15 The Right to Water' (2002)

Conti KI and Gupta J, 'Global Governance Principles for the Sustainable Development of Groundwater Resources' (2016) 16 International Environmental Agreements: Politics, Law and Economics 849

Cosens B and others, 'Identifying Legal, Ecological and Governance Obstacles, and Opportunities for Adapting to Climate Change' (2014) 6 Sustainability 2338 <www.mdpi.com/2071-1050/6/4/2338/> accessed 25 April 2014

Cullet P and others, *Water Law for the Twenty-First Century National and International Aspects of Water Law Reform in India* (2010)

Curry E, 'Water Scarcity and the Recognition of the Human Right to Safe Freshwater' (2010) 9 Northwestern Journal of International Human Rights 103

Curry E, 'Water Scarcity and the Recognition of the Human Right to Safe Freshwater' (2010) 9 Northwestern Journal of International Human Rights 103

Daci J, 'Protection of the Human Right to Water Under International Law The Need for a New Legal Framework' (2012) 6 Academicus International Scientific Journal 71

Dai L, 'Identifying and Understanding the Main Challenges for Sustainable Water Resource Management in China' (2014) 24 Journal of Water Law 249

Dai L, 'A New Perspective on Water Governance in China: Captain of the River' (2015) 40 Water International 87 <www.tandfonline.com/doi/abs/10.1080/02508060.2014.986702>

Davis AN, 'Intersectionality and International Law: Recognizing Complex Identities on the Global Stage' (2015) 28 Harvards Human Rights Journal 205

Davis SH, 'Indigenous Peoples and Climate Change' (2010) 1 The International Indigenous Policy Journal 1

De Albuquerque C, 'Climate Change and the Human Rights to Water' (2010) <www.ohchr.org/Documents/Issues/Water/Climate_Change_Right_Water_Sanitation.pdf>

De Albuquerque C, 'Report of the Independent Expert on the Issue of Human Rights Obligations Related to Access to Safe Drinking Water and Sanitation "Compilation of Good Practices"' (2010)

De Albuquerque C, 'Report of the Special Rapporteur on the Human Right to Safe Drinking Water and Sanitation' (2013)

De Albuquerque C and Roaf V, *On The Right Track. Good Practices in Realising the Rights to Water and Sanitation* (2012th edn, 2012) <www.ohchr.org/Documents/Issues/Water/BookonGoodPractices_en.pdf>

De Albuquerque C and Roaf V, 'Financing', *Realising the human rights to water and sanitation: A handbook by the UN Special Rapporteur Catarina De Albuquerque* (2014)

De Albuquerque C and Roaf V, 'Introduction' in Virginia Roaf, Inga Winkler and Muriel Schiessl (eds), *Realising the human rights to water and sanitation: A handbook by the UN Special Rapporteur Catarina De Albuquerque* (2014)

De Albuquerque C and Roaf V, 'Monitoring', *Realising the human rights to water and sanitation: A handbook by the UN Special Rapporteur Catarina De Albuquerque* (2014)
De Albuquerque C and Roaf V, 'Principles', *Realising the human rights to water and sanitation: A handbook by the UN Special Rapporteur Catarina De Albuquerque* (2014)
De Albuquerque C and Roaf V,'Services', *Realising the human rights to water and sanitation: A handbook by the UN Special Rapporteur Catarina De Albuquerque* (2014)
De Albuquerque C and Roaf V, 'Sources', *Realising the human rights to water and sanitation: A handbook by the UN Special Rapporteur Catarina De Albuquerque* (2014)
de Theije M, ' 'De Brazilianen Stelen Al Ons Goud!' Braziliaanse Migranten in Stad En Binnenland' (2007) 26 OSO, Tijdschrift voor Surinamistiek en het Caribische Gebied 81
De Vido S, 'The Right to Water as an International Custom : The Implications in Climate Change Adaptation Measures' (2012) 3 Carbon and Climate Law Review 221
de Vries E, *Suriname Na de Binnenlandse Oorlog* (LM Publishers 2005)
del Prado N, 'Naar Geintegreerd Waterbeheer in Suriname' (2011)
del Prado N, 'Analysis of Instruments, Principles and Legal Norms in Force on Domestic Law of the Amazon Basin in Water Resources Management' (2013)
Del Prado N, 'An Analysis of Land Rights of the Indigenous Peoples and Maroons in Suriname. Adaptation of Legislation in Suriname' (2006)
Deleuil T, 'The Common but Differentiated Responsibilities Principle: Changes in Continuity after the Durban Conference of the Parties' (2012) 21 Review of European Community and International Environmental Law 271
Dellapenna JW, 'Is Sustainable Development a Serviceable Legal Standard in the Management of Water?' [2004] Water Recources Update 87
Dellapenna JW, 'The Customary Law Applicable to Internationally Shared Groundwater' (2011) 36 Water International 584 <www.tandfonline.com/doi/abs/10.1080/02508060.2011.599025>
Dellapenna JW and Gupta J, 'The Evolution of Global Water Law' in Joseph W. Dellapenna; Joyeeta Gupta; (ed), *The Evolution of the Law and Politics of Water* (2009th edn, Springer Netherlands 2009)
Dennis MJ and Stewart DP, 'Justiciability of Economic, Social, and Cultural Rights : Should There Be an International Complaints Mechanism to Adjudicate the Rights to Food, Water, Housing, and Health ?' (2012) 98 The American Journal of International Law 462
Dernbach JC and Mintz J a., 'Environmental Laws and Sustainability: An Introduction' (2011) 3 Sustainability 531 <www.mdpi.com/2071-1050/3/3/531/> accessed 12 February 2014
Desmet E, 'The Twin of Green The Rights of Indigenous Peoples and Local Communities Entwined with Nature Conservation' (Katholieke Universiteit Leuven 2010) <https://lirias.kuleuven.be/handle/123456789/259914>
Desmet E, *Indigenous Rights Entwined with Nature Conservation* (Intersentia 2011)
Dinokopila BR, 'The Right to Water in Botswana : A Review of the Matsipane' (2014) 11 African Human Rights Law Journal 282
Dinstein Y, 'The Right to Life, Physical Integrity, and Liberty' in Louis Henkin (ed), *The International Bill of Rights. The Covenant on Civil and Political Rights* (Columbia University Press 1981)
Donoho D, 'Some Critical Thinking about a Human Right to Water' (2012) 19 ILSA Journal of International and Comparative Law 91
Drexhage J and Murphy D, 'Sustainable Development: From Brundtland to Rio 2012' (2010)

Duwell M, 'Human Dignity and Future Generations' in Marcus Duwell and others (eds), *The Cambridge Handbook of Human Dignity Interdisciplinary Perspectives* (Cambridge University Press 2014)

Edenhofer O and others, *Climate Change, Justice and Sustainability: Linking Climate and Development Policy* (Springer 2012) <http://link.springer.com.ezproxy.usherbrooke.ca/chapter/10.1007/978-94-007-4540-7_5>

Ellis K and Feris L, 'The Right to Sanitation: Time to Delink from the Right to Water' (2014) 36 Human Rights Quarterly 607 <http://muse.jhu.edu/content/crossref/journals/human_rights_quarterly/v036/36.3.ellis.html> accessed 11 August 2014

Ellis K and Feris L, 'The Right to Sanitation: Time to Delink from the Right to Water' (2014) 36 Human Rights Quarterly 607 <http://muse.jhu.edu/content/crossref/journals/human_rights_quarterly/v036/36.3.ellis.html>

Ellison K, 'Rio+20: How the Tension Between Developing and Developed Countries Influenced Sustainable Development Efforts' (2014) 27 Global Business & Development Law Journal 107

Feitelson E and Chenoweth J, 'Water Poverty: Towards a Meaningful Indicator' (2002) 4 Water Policy 263 <http://linkinghub.elsevier.com/retrieve/pii/S1366701702000296>

Fernández-Jáuregi C and Crespo Milliet A, 'Water, a Unique Resource' in Mikel Mancisidor and Natalie Uribe (eds), *The Human Right to Water: Current Situation and Future Challenges* (UNESCO 2008)

Filcak R, 'On the Distribution of the Environment Benefits and Adverse Impacts: The Environmental Justice and Vulnerability of People in Risk' (2009) 1 Environmental Law Review of Eastern & Central Europe 63 <www.prog.sav.sk/subory/esee/8_vulnerability_paper.pdf>

Filmer-Wilson E, 'The Human Rights-Based Approach to Development: The Right to Water' (2005) 23 Netherlands Quarterly of Human Rights 213

Fineman MA, 'Equality, Autonomy, and the Vulnerable Subject in Law and Politics' in Martha Albertson Fineman and Anna Grear (eds), *Vulnerability: reflections on a new ethical foundation for law and politics* (2nd edn, Routledge 2016)

Fischer-Lescano A and Teubner G, 'Regime-Collisions: The Vain Search for Legal Unity in the Fragmentation of Global Law' (2004) 25 Michigan Journal of International Law 999

Fisher D, *The Law and Governance of Water Resources* (Edward Elgar Publishing 2009)

Fitzmaurice M, 'The Human Right to Water' (2006) 18 Fordham Environmental Law Review 537

Fons Coomans, *Justiciability of Economic and Social Rights: Experiences from Domestic Systems* (Intersentia 2006)

Food and Agriculture Organization of the United Nations, *Land Resources Information Systems in the Caribbean* (2001)

Freestone D and Hey E (eds), *The Precautionary Principle and International Law : The Challenge of Implementation* (Kluwer Law International 1996)

French D and Cooper NJ, 'The Right to Water in South Africa: Constitutional Managerialism and a Call for Pluralism' (2012) 5 <http://eprints.lincoln.ac.uk/6738/>

Fukuda-parr S and Lawson-remer T, 'Measuring the Progressive Realization of Human Rights Obligations : An Index of Economic and Social Rights Fulfillment Department of Economics Working Paper Series' (2008) Paper 2008 Economics Working Papers 1

Garcia B, 'The 1978 Amazon Cooperation Treaty', *The Amazon from an International Law Perspective*, vol 1 (Cambridge University Press 2011)

Gavouneli M, 'A Human Right to Groundwater?' (2011) 13 International Community Law Review 305 <http://ezp-prod1.hul.harvard.edu/login?url=http://search.ebscohost.com/login.aspx?direct=true&db=aph&AN=63526103&site=ehost-live&scope=site&scope=cite> accessed 8 March 2013

Getches DH, 'Indigenous Peoples' Rights to Water Under International Norms' (2005) 16 Colorado Journal of International Environmental Law and Policy 259

Giorgetta S, 'The Right to a Healthy Environment, Human Rights and Sustainable Development' (2002) 2 International Environmental Agreements: Politics, Law and Economics 173 <http://link.springer.com/article/10.1023/A:1020938009559>

Gleick PH, *California Water 2020: A Sustainable Vision* (Pacific Institute for Studies in Development, Environment, and Security 1995)

Gleick PH, 'The Human Right to Water' (1998) 1 Water Policy 487

Gleick PH, 'WATER IN CRISIS : PATHS TO SUSTAINABLE WATER USE' (1998) 8 Ecological Applications 571

Goepel M, 'Formulating Future Just Policies: Applying the Delhi Sustainable Development Law Principles' (2010) 2 Sustainability 1694 <www.mdpi.com/2071-1050/2/6/1694/> accessed 12 February 2014

Government of Suriname, 'MDG Progress Report 2009' (2009)

Government of Suriname and Inter-American Development Bank, 'Suriname Water Supply Infrastructure Rehabilitation' (2010)

Government of Suriname and United Nations Country Team Suriname, 'Suriname MDG Baseline Report' (2005)

Government of the Republic of Suriname, 'National Report in Preparation of The Third International Conference on Small Island Developing States (SIDS)' (2013)

Government of the Republic of Suriname, 'MGD Progress Report 2014' (2014)

'Governments Contribution to the Implementation of the Sustainable Development Goals Related to Water | 2015 UN-Water Annual International Zaragoza Conference. Water and Sustainable Development: From Vision to Action. 15–17 January 2015' <www.un.org/waterforlifedecade/waterandsustainabledevelopment2015/stakeholders_governments_and_local_authorities.shtml> accessed 5 February 2016

Grafton RQ and others, *Global Water: Issues and Insights* (ANU Press 2014)

Grant A, 'International Court and Environmental Human Rights: Re-Imagining Adjudicative Paradigms' in Anna Grear and Louis J Kotze (eds), *Research Handbook and the Environment* (Edward Elgar Publishing 2015)

Greenfield SMR, 'Challenging the Rhetoric of Sustainability and Participation a Case Study of the Ghana Upper Region Water Supply Programme.pdf' (Carleton University 1990)

Groce N and others, 'Water and Sanitation Issues for Persons with Disabilities in Low- and Middle-Income Countries: A Literature Review and Discussion of Implications for Global Health and International Development' (2011) 9 Journal of water and health 617

Groce N and others, 'Water and Sanitation Issues for Persons with Disabilities in Low- and Middle-Income Countries: A Literature Review and Discussion of Implications for Global Health and International Development.' (2011) 9 Journal of water and health 617 <www.ncbi.nlm.nih.gov/pubmed/22048421> accessed 7 April 2014

Gupta J, Ahlers R and Ahmed L, 'The Human Right to Water : Moving Towards Consensus in a Fragmented World' (2010) 19 Review of European Community and International Environmental Law 294

Gupta J and Dellapenna JW, 'The Challenges for the Twenty-First Century : A Critical Approach' in Joseph W; Dellapenna and Joyeeta; Gupta (eds), *The Evolution of the Law and Politics of Water* (2009th edn, Springer Science 2009)

Hadjiargyrou Z, 'A Conceptual and Practical Evaluation of Intergenerational Equity in International Environmental Law' (2016) 18 International community law review 248

Haidera M and others, 'Water Scarcity and Climate Change Adaptation for Yemen's Vulnerable Communities' (2011) 16 Local Environment 473 <www.tandfonline.com/doi/abs/10.1080/13549839.2011.565465> accessed 6 November 2013

Hall RP, Van Koppen B and Van Houweling E, 'The Human Right to Water: The Importance of Domestic and Productive Water Rights' (2013) 20 Science and engineering ethics <www.ncbi.nlm.nih.gov/pubmed/24337891> accessed 25 July 2014

'Handleiding Burgerparticipatie En Communicatie Voor de Regionale Organen' <http://basahmadali.nl/wp-content/uploads/2015/04/Handleiding-Burgerparticipatie-en-Communicatie-Districtsraad-Paramaribo-06-07-2012.pdf>

Hannon KL and Lawrence RS, 'Antropogenic Sources of Water Pollution: Parts 1 and 2' in Janine MH Selendy (ed), *Water and Sanitation-Related Diseases and the Environment: Challenges, Interventions, and Preventive Measures* (First, John Wiley & Sons, Inc 2011)

Hardberger A, 'Life, Liberty, and the Pursuit of Water : Evaluating Water as a Human Right and the Duties and Obligations It Creates' (2005) 4 Northwestern Journal of International Human Rights 331

Harding A, *Access to Environmental Justice: A Comparative Study* (Nijhoff 2007)

Harnum J, 'Deriving the Right to Water from the Right to Life, Liberty and Security of the Person : Section 7 of the Canadian Charter of Rights and Freedoms and Aboriginal Communities in Canada' (2010) 19 Review of European Community and International Environmental Law 306

Harris LM, 'Water Rich, Resource Poor: Intersections of Gender, Poverty, and Vulnerability in Newly Irrigated Areas of Southeastern Turkey' (2008) 36 World Development 1 <http://linkinghub.elsevier.com/retrieve/pii/S0305750X08001319> accessed 18 November 2013

Heemskerk M, 'Livelihood Decision Making and Environmental Degradation: Small-Scale Gold Mining in the Suriname Amazon' (2002) 15 Society & Natural Resources 327

Heemskerk M and Kooye R Van Der, 'Challenges To Sustainable Small-Scale Mine Development in Suriname' in Gavin Hilson (ed), *The Socio-Economic Impacts of Artisanal and Small-Scale Mining in Developing Countries* (Taylor and Francis 2003)

Heenan J, Fadaei H and Walker S, 'Sustainable Development' [2002] NGLS Roundup 1

Hellum A, Kameri-Mbote P and Van Koppen B, *Water Is Life* (2015)

Hennebel L, 'The Inter-American Court of Human Rights: The Ambassador of Universalism' (2011) Special ed Quebec Journal of International Law 57

Hermans L, 'Climate Change, Water Stress, Conflict and Migration. Taking Stock of Current Insights through a Vulnerability Lens', *Climate change, water stress, conflict and migration* (2012)

Hildering A, *International Law, Sustainable Development and Water Management* (1st edn, Eburon Uitgeverij BV 2004)

Hobson C, 'Human Security after the Shock: Vulnerability and Empowerment' in Christopher Hobson, Paul Bacon and Robin Cameron (eds), *Human Security and Natural Disasters* (Routledge/ Taylor & Francis 2014) <http://isp.unu.edu/research/human-security/index.html>

Hoefte R, *Suriname in the Long Twentieth Century: Domination, Contestation, Globalization* (2013) <http://books.google.com/books?id=sHdanQEACAAJ&pgis=1>
Hoever-Venoaks MR and Damen LJA, *Surinaams Bestuursrecht* (Ars Aequi Libri 2003)
Howard G and Bartram J, 'Domestic Water Quantity, Service Level and Health' (2002)
Howley J, 'The Gabcikovo Nagymaros Case: The Influence of the International Court of Justice on the Law of Sustainable Development' (2009) 2 Queensland Law Student Review 1
'Human Rights and the Environment',, *Proceedings of the 13th Informal ASEM Seminar on Human Rights* (Asia-Europe Foundation (ASEF) 2013) <www.pubmedcentral.nih.gov/articlerender.fcgi?artid=3638120&tool=pmcentrez&rendertype=abstract>
Human Rights Council, 'Report of the Special Rapporteur on the Human Right to Safe Drinking Water and Sanitation, Catarina De Albuquerque Mission to Senegal (14 to 21 November 2011)'
Humphreys S, *Human Rights and Climate Change* (Stephen Humphreys ed, Cambridge University Press 2009)
Hunt P and Backman G, 'Health Systems and the Right to the Highest Attainable Standard of Health', *Human Rights and Climate Change* (Cambridge University Press 2009)
International Development Bank, 'Governance in Suriname' (2001)
International Law Association, 'ILA New Delhi Declaration of Principles of International Law Relating to Sustainable Development' (2002) 2 International Environmental Agreements: Politics, Law and Economics 211
Ippolito F and Iglesias Sánchez S, 'Introduction' in Francesca Ippolito and Sara Iglesias Sánchez (eds), *Protecting Vulnerable Groups. The European Human Rights Framework* (Hart Publishing 2015)
Iribarnegaray M and Seghezzo L, 'Governance, Sustainability and Decision Making in Water and Sanitation Management Systems' (2012) 4 Sustainability 2922 <www.mdpi.com/2071-1050/4/11/2922/> accessed 26 February 2014
Janssen R, *In Search of a Path An Analysis of the Foreign Policy of Suriname from 1975 to 1991* (KITLV Press 2011)
Jefferords C and Shah F, 'On the Natural and Economic Difficulties to Fulfilling the Human Right to Water Within a Neoclassical Economics Framework' (2012)
JMP, World Health Organization and UNICEF, 'Ending Inequalities – a Cornerstone of the Post-2015 Development Agenda' 1
Jóhannsdóttir A, Cresswell I and Bridgewater P, 'The Current Framework for International Governance of Biodiversity : Is It Doing More Harm Than Good ?' (2010) 19 Review of European Community and International Environmental Law 139
Kambel E-R and MacKay F, *De Rechten van Inheemse Volken En Marrons in Suriname* (KITLV Uitgeverij 2003)
Kambel ER, 'Mijnbouw, Infrastructuur En de Rechten van Inheemse En Tribale Volkeren in Zuid-Oost Suriname' [2009] Project strategische analse en participatief actieplan voor zuid oost Suriname
Khalfan A and others, *Manual on the Right to Water and Sanitation* (2008)
Kiefer T and others, *Legal Resources for the Right to Water and Sanitation* (2nd edn, Centre on Housing Rights and Evictions 2008) <www.cohre.org/water>
Kiefer T and others, *The Human Right to Safe Drinking Water and Sanitation in Law and Policy – a Sourcebook* (WASH United, Freshwater Action Network, Waterlex 2012)

Kiefer T and Brölmann C, 'Beyond State Sovereignty: The Human Right to Water' (2005) 5 Non-State Actors and International Law 183 <www.ingentaselect.com/rpsv/cgi-bin/cgi?ini=xref&body=linker&reqdoi=10.1163/157180705775435465>

Kirby M, 'International Law – The Impact on National Constitutions' (2006) 21 American University International Law Review 327

Kirchmeier F, 'The Right to Development – Where Do We Stand ? State of the Debate on the Right to Development Dialogue on Globalization', vol 23 (2006)

Klug F, *A Magna Carta for All Humanity*, vol 226 (Routledge 2015)

Ko J and Donahue WF, *Moving Waters : Water Management Options to Achieve Social, Economic, and Environmental Goals* (Water Matters Society of Alberta 2012)

Kole H, 'Het Poldermodel. Politieke Cultuur in de Waterschappen (c.1400-c.1800).' (Utrecht University 2017)

Kotze LJ, 'Human Rights, the Environmental, and the Global South', *International Environmental Law and the Global South* (Cambridge University Press 2015)

Krchnak KM, 'Improving Water Governance Through Increased Public Access to Information and Participation' (2004) 5 Sustainable Development Law & Policy 34

Langford M, 'The United Nations Concept of Water as a Human Right: A New Paradigm for Old Problems?' (2005) 21 International Journal of Water Resources Development 273 <www.tandfonline.com/doi/abs/10.1080/07900620500035887> accessed 6 March 2013

Langford M, 'Ambition That Overleaps Itself? A Response to Stephen Tully's Critique of the General Comment on the Right to Water' (2006) 24 Netherlands Quarterly of Human Rights 433

Larson RB, 'Water, Worship, and Wisdom: Indigenous Traditional Ecological Knowledge and the Human Right to Water' (2012) 19 ILSA Journal of International and Comparative Law 43

Lindhout PE and Den Broek B van, 'The Polluter Pays Principle: Guidelines for Cost Recovery and Burden Sharing in the Case Law of the European Court of Justice' (2014) 10 Utrecht Law Review 46

Maggio GF, 'Inter/Intra-Generational Equity : Current Applications under International Law for Promoting the Sustainable Development of Natural Resources' 4 Buffalo environmental law journal 161

Marong ABM, 'From Rio to Johannesburg: Reflections on the Role of International Legal Norms in Sustainable Development' (2003) 16 Georgetown International Environmental Law Review 21

Matsui Y, 'Some Aspects of the Principle of " Common but Differentiated Responsibilities "' (2002) 2 International Environmental Agreements: Politics, Law and Economics 151

May JR and Daly, 'Emerging Environmental Constitutionalism', *Global Environmental Constitutionalism* (Cambridge University Press 2015)

May JR and Daly E, 'Water and Environmental Constitutionalism', *Global Environmental Change* (Cambridge University Press 2015)

Mccaffrey SC, 'Keynote: Sustainability and Sovereignty in the 21th Century' (2013) 41 Denver Journal of International Law and Policy 507

Mccaffrey SC, 'Human Right to Water: A False Promise' (2016) 47 University of the Pacific Law Review 221

McCaffrey SC, 'A Human Right to Water : Domestic and International Implication' (1992) 5 Georgetown International Environmental Law Review 1

Mcintyre O, 'The Role of Customary Rules and Principles of International Environmental Law in the Protection of Shared International Freshwater Resources' (2006) 46 Natural Resources Journal 158

Mearns R and Norton A, 'Equity and Vulnerability in a Warming World: Introduction and Overview' in Robin Mearns and Andrew Norton (eds), *The Social Dimensions of Climate Change Equity and Vulnerability in A Warming World* (The World Bank 2010)

Melo MP, 'Efficiency, Sustainability, and Justice to Future Generations' in Klaus Mathis (ed), *Efficiency, Sustainability, and Justice to Future Generations*, vol 98 (Springer Netherlands 2011) <http://link.springer.com/10.1007/978-94-007-1869-2> accessed 18 September 2013

Menke J (ed), *Mozaiek van Het Surinaamse Volk* (2016)

Meteorological Service Suriname, 'Flash Flood Guidance' 1 <www.wmo.int/pages/prog/hwrp/documents/FFI/Flash_Flood_Guidance_Systems_Background_Information.pdf>

'Milieustatistieken Environment Statistics'

Ministry of Labour Technological Development and Environment, 'Second National Communication to the United Nations Framework Convention on Climate Change' (2013) <www.nimos.org/smartcms/downloads/1. SURINAME SNC-UNFC on ClimateChange February 2013.pdf>

Mirosa O and Harris LM, 'Human Right to Water: Contemporary Challenges and Contours of a Global Debate' (2012) 44 Antipode 932 <http://doi.wiley.com/10.1111/j.1467-8330.2011.00929.x> accessed 6 March 2013

Mirumachi N, 'Water and Sustainable Development' in Michael Redclift (ed), *Routledge International Handbook of Sustainable Development* (Taylor and Francis 2015) <www.un.org/waterforlifedecade/waterandsustainabledevelopment2015/images/sustainable_development_eng.pdf>

Misiedjan D, 'Ten Building Blocks for a Sustainable Human Right to Water: An Integrated Method to Assess Sustainable Access to Water for Domestic Purposes of Vulnerable Groups' in Erin Daly and others (eds), *New Frontiers in Environmental Constitutionalism* (UNEP 2017)

Misiedjan D and Gupta J, 'Indigenous Communities: Analyzing Their Right to Water under Different International Legal Regimes' (2014) 10 Utrecht Law Review 77

Misiedjan D, van Rijswick M and Tjen A Kwoei A, 'A Human Right to Water While the Well Runs Dry: Analysing the Legal and Regulatory Framework of Yemen Water Law' (2015) 24 The Journal of Water Law 199

Morawa AHE, 'Vulnerability as a Human Rights Law' (2003) 6 139

Morris JDK and Ruru J, 'Giving Voice to Rivers: Legal Personality as a Vehicle for Recognizing Indigenous Peoples' Relationship to Water?' (2011) 14 Australian Indigenous Law Review 49

Muller B, 'Varieties of Distributive Justice in Climate Change' (2001) 48 Climate Change 273 <http://jmq.sagepub.com/lookup/doi/10.1177/107769900908600101>

Murillo Chávarro J, 'The Right to Water in the Case-Law of the Inter-American Court of Human Rights' (2014) 7 39

Murillo Chávarro J, *The Human Right to Water A Legal Comparative Perspective at the International, Regional and Domestic Level* (Intersentia 2015)

N.V. Surinaamse Waterleiding Maatschappij, 'Masterplan for the Coastal Zone' (2002)

N.V. Surinaamse Waterleiding Maatschappij, 'Voorwaarden Voor Aansluiting En Waterlevering' 1

N.V. Surinaamse Waterleiding Maatschappij, 'Suriname Water Supply Master Plan'

N.V. Surinaamse Waterleiding Maatschappij, 'Informatiebrochure SWM' 1
N.V. Surinaamse Waterleiding Maatschappij, 'Drinkwatertarieven'
NIMOS, 'Environmental and Social Analysis/environmental and Social Management Framework' (2010)
Noga J and Wolbring G, 'The Economic and Social Benefits and the Barriers of Providing People with Disabilities Accessible Clean Water and Sanitation' (2012) 4 Sustainability 3023
Nurmohamed RJ, 'The Impact of Climate Change and Climate Variability on Th Water Resources in Suriname' (Anton de Kom Universiteit van Suriname 2008)
O'Reilly K, Dhanju R and Goel A, 'Exploring "The Remote" and "The Rural": Open Defecation and Latrine Use in Uttarakhand, India' (2017) 93 World Development 193
Obani P and Gupta J, 'The Evolution of the Right to Water and Sanitation: Differentiating the Implications' (2015) 24 Review of European, Comparative and International Environmental Law 27
Obani P and Gupta J, 'Human Right to Sanitation in the Legal and Non-Legal Literature: The Need for Greater Synergy' (2016) 3 Wiley Interdisciplinary Reviews: Water <http://doi.wiley.com/10.1002/wat2.1162>
Odello M and Seatzu F, 'The International Covenant on Economic, Social and Cultural Rights', *The UN Committee on economic, social and cultural rights: the law, process and practice* (Routledge 2013)
OECD, 'OECD Principles on Water Governance' (2015)
Office of the High Commissioner, *Human Rights Indicators A Guide to Measurement* (2012)
Ofosu-Mensah EA, 'Historical Overview of Traditional and Modern Gold Mining in Ghana' (2011) 1 International Research Journal of Library, Information and Archival Studies 6
OHCHR, *Fact Sheet No. 35, The Right to Water* (2010) <www.ohchr.org/Documents/Publications/FactSheet35en.pdf>
Onzivu W, 'The Long Road to Integrating Public Health into Sustainable Development of Shared Freshwaters in International Environmental Law: Lessons from Lake Victoria in East Africa' (2012) 46 International Lawyer 867
Orebech P and Chanock M, 'Towards Sustainability : The Basis in International Law' in Fred Bosselman and others (eds), *The Role of Customary Law in Sustainable Development* (Cambridge University Press 2006)
Oswald Spring Ú, 'Social and Environmental Vulnerability in a River Basin of Mexico' in Úrsula Oswald Spring, Hans Günter Brauch and Keith G Tidball (eds), *Expanding Peace Ecology: Peace, Security, Sustainability, Equity and Gender*, vol 12 (Springer International Publishing 2014) <http://link.springer.com/10.1007/978-3-319-00729-8> accessed 22 December 2013
Ouboter PE and others, 'Mercury Levels in Pristine and Gold Mining Impacted Aquatic Ecosystems of Suriname, South America' (2012) 41 Ambio 873
Padilla E, 'Intergenerational Equity and Sustainability' (2002) 41 Ecological Economics 69
PAHO/WHO, 'PAHO/WHO Country Cooperation Strategy Suriname 2012 – 2016' (2012)
Palmer N, 'Water Adequacy and Essential Element for Empowering People' (2012) 19 ILSA Journal of International and Comparative Law 31
Palmer N, 'Water Adequacy as an Essential Element for Empowering People' (2012) 19 ILSA Journal of International and Comparative Law 31
Pejan R, 'The Right to Water: The Road to Justiciability' (2004) 36 The George Washington International Law Review 1181

Peplow D and Augustine S, 'Community-Led Assessment of Risk from Exposure to Mercury by Native Amerindian Wayana in Southeast Suriname' (2012) 2012 Journal of Environmental and Public Health 1

Peroni L and Timmer A, 'Vulnerable Groups: The Promise of an Emerging Concept in European Human Rights Convention Law' (2013) 11 International Journal of Constitutional Law 1056

Pienaar H, Belcher A and Grobler DF, 'Protecting Aquatic Ecosystem Health for Sustainable Use' in Barbara Schreiner and Rashid Hassan (eds), *Transforming Water Management in South Africa: Designing and Implementing a New Policy Framework* (Springer Science 2011) <http://link.springer.com/10.1007/978-90-481-9367-7>

Pieterse M, 'Coming to Terms with Judicial Enforcement of Socio-Economic Rights' (2004) 20 South African Journal on Human Rights 383

Planbureau Suriname, 'Regionaal Plan Brokopondo 2015 – 2040' (2013)

Ramlal V, 'De Huisvestingsproblematiek in Suriname' (Anton de Kom Universiteit van Suriname 2013)

Ravenberg R, 'Water Management in Suriname' (2013)

Rayner S and Malone EL, 'Climate Change, Poverty, and Intragenerational Equity: The National Level' (2001) 1 International Journal of Global Environmental Issues 175

Razzaque J, 'Public Participation in Water Governance' in Joseph W Dellapenna and Joyeeta Gupta (eds) (Springer Netherlands 2009)

Regering van de Republiek Suriname, 'Ontwikkelingsplan 2012 – 2016 Suriname in Transformatie' (2012)

Ribot J, 'Vulnerability Does Not Fall from the Sky: Toward Multiscale, Pro-Poor Climate Policy' in Andrew Norton and Robin Mearns (eds), *Social Dimensions of Climate Change: Equity and Vulnerability in a Warming World* (The World Bank 2010)

Richard P. Hiskes, *The Human Right to a Green Future* (Cambridge University Press 2009)

Riedel E, 'The Human Right to Water and General Comment No. 15 of the CESCR' in Eibe Riedel and Peter Rothen (eds), *The Human Right to Water* (Berlines Wissenschafts – Verslag 2006)

Riedel E, Giacca G and Golay C, *Economic, Social, and Cultural Rights in International Law* (Eibe Riedel, Gilles Giacca and Christophe Golay eds, Oxford University Press 2014)

Riedel E, Giacca G and Golay C, 'The Development of Economic, Social, and Cultural Rights in International Law', *Economic, Social, and Cultural Rights in International Law* (2014)

Rights. RLC of H,) IC of J (1952- and Rights. UMI for H, 'The Maastricht Guidelines on Violations of Economic, Social and Cultural Rights' (1998) 20 Human Rights Quarterly 691 <http://muse.jhu.edu/journals/human_rights_quarterly/v020/20.3intl_comm.html> accessed 4 February 2016

Roaf BV, Khalfan A and Langford M, 'Monitoring Implementation of the Right to Water: A Framework for Developing Indicators' (2005) 14 <www.worldwatercouncil.org/fileadmin/wwc/Programs/Right_to_Water/Pdf_doct/Monitoring_implementation_of_the_RTW_Indicators.pdf>

Robertson RE, 'Measuring State Compliance with the Obligation to Devote the " Maximum Available Resources " to Realizing Economic, Social, and Cultural Rights' (1994) 16 Human Rights Quarterly 693

Robesin MA, 'Katoen En Waterschaarste', *Duurzame handel in juridisch perspectief* (2014)

Robinson NA, 'Reflecting on Rio: Environmetal Law in the Coming Decades' in Jamie Benidickson (ed), *Environmental Law and Sustainability after Rio* (Edward Elgar Publishing 2011)

Salman M. A. Salman; Siobhán McInerney-Lankford, *The Human Right to Water* (David Freestone; Salman MA Salman; ed, 2004th edn, The International Bank for Reconstruction and Development / The World Bank 2004)

Salomon M and Sengupta A, 'The Right to Development: Obligations of States and the Rights of Minorities and Indigenous Peoples' (2003)

Sancin V and Dine MK, 'Ensuring Access to Safe Drinking Water as an Imperative of Sustainable Development' in Volker Mauerhofer (ed), *Legal Aspects of Sustainable Development: Horizontal and Sectorial Policy Issues* (Springer International Publishing Switserland 2016)

Sandie G, 'Sunny Point Access to Water as a Human Right in the Coastal Region of Suriname' (FHR Lim A Po Institute 2014)

Sarkin J and Cook A, 'The Human Rights of the San (Bushmen) of Botswana – the Clash of the Rights of Indigenous Communities and Their Access to Water with the Rights of the State to Environmental Conservation and Mineral Resource Exploitation' (2011) 20 Journal of International Law and Policy 1

Satterthwaite ML and Simeone JC, 'A Conceptual Roadmap for Social Science Methods in Human Rights Fact-Finding' in Philip Alston and Sarah Knuckey (eds), *The Transformation of Human Rights Fact-Finding* (Oxford University Press 2016)

Scanlon J, Cassar A and Nemes N, *Water as a Human Right?* (IUCN 2004)

Schlosser C and others, 'The Future of Global Water Stress: An Integrated Assessment', vol 2 (2014) <http://onlinelibrary.wiley.com/doi/10.1002/2014EF000238/abstract>

Schmidt-Traub G, 'The Millennium Development Goals and Human Rights-Based Approaches: Moving towards a Shared Approach' (2009) 13 The International Journal of Human Rights 72 <www.tandfonline.com/doi/abs/10.1080/13642980802532374> accessed 6 January 2015

Schrijver N, 'Conclusions: The Challenges to Sustainable Development and to Its Further Evolution in International Law', *The Evolution of Sustainable Development in International Law : Inception, Meaning and Status* (2008)

Schrijver N, *The Evolution of Sustainable Development in International Law: Inception, Meaning and Status* (Martinus Nijhof Publishers 2008)

Schwartz P, 'Principle 16: The Polluter-Pays Principle' in Jorge E Vinuales (ed), *The Rio Declaration on Environment and Development : a Commentary* (Oxford University Press 2015)

Shelton D, 'Human Rights, Environmental Rights, and the Right to Environment' (1991) 28 Stanford Journal of International Law 103

Shelton D, 'The Environmental Jurisprudence of International Human Rights Tribunals' in Romina Picolotti and Jorge Daniel Taillant (eds), *Linking Human Rights and the Environment* (The University of Arizona Press 2003)

Shelton D, 'Intergenerational Equity' in Rudiger Wolfrum and Chie Kojima (eds), *Solidarity: a Structural Principle of International Law* (Springer Berlin Heidelberg 2010)

Shelton D and Kiss A, *Judicial Handbook on Environmental Law* (UNEP 2005) <www.unep.org/delc/Portals/119/publications/Judicial-Handbook-Environmenal-Law.pdf>

Sijniensky RI, 'From the Non-Discrimination Clause to the Concept of Vulnerability in International Human Rights Law' in Yves Haeck and others (eds), *The Realisation of Human Rights: When Theory Meets Practice* (Intersentia 2014)

Simma B and Alston P, 'The Sources of Human Rights Law: Custom, Jus Cogens, and General Principles' [1988] Australian Yearbook of Law 82

Singh N and others, 'Accessing Water through a Rights-Based Approach: Problems and Prospects Regarding Children' (2012) 14 Water Policy 298 <www.iwaponline.com/wp/01402/wp014020298.htm> accessed 11 March 2013

Small Arms Survey, 'Under Pressure: Social Violence over Land and Water in Yemen' (2010) <www.genevadeclaration.org/fileadmin/docs/regional-publications/Social_violence_over_land_and_water_in_Yemen.pdf>

Smit B and Wandel J, 'Adaptation, Adaptive Capacity and Vulnerability' (2006) 16 Global Environmental Change 282 <http://linkinghub.elsevier.com/retrieve/pii/S0959378006000410> accessed 19 March 2014

Sobsey MD, 'Managing Water in the Home: Accelerated Health Gains from Improved Water Supply' (2004) 8 World Health 1 <www.bvsde.paho.org/bvsacd/who/sobs.pdf>

Speed, R. Yuanyuan, L. Le Quesne, T. Pegram, G. Zhiwei Z, *Basin Water Allocation Planning: Principles, Procedures and Approaches for Basin Allocation Planning* (UNESCO 2013)

Spring ÚO and Brauch HG, 'Securitizing Water' (2009) 4 Facing Global Environmental Change 175 <http://dx.doi.org/10.1007/978-3-540-68488-6_11>

Ssenyonjo M, *Economic, Social and Cultural Rights in International Law* (2nd edn, Hart Publishing 2016)

Stenekes N, 'Sustainability and Participation in the Governing of Water Use: The Case of Urban Water Recycling' (University of New South Wales 2006)

Sultana F, 'Water, Water Everywhere, But Not a Drop to Drink: Pani Politics (Water Politics) in Rural Bangladesh' (2007) 9 International Feminist Journal of Politics 494 <www.tandfonline.com/doi/abs/10.1080/14616740701607994>

Taher T and others, 'Informing and Enabling Local Ground Water Governance for Yemen' (2011)

Tarlock AD (ed), *Environmental Laws and Their Enforcement* (EOLSS Publishers Co Ltd 2009)

The Ministry of Labour Technological Development and Environment, 'The Fourth National Report to the Convention on Biological Diversity' (2013)

The World Bank and others, *Sanitation for Development* (Second edi, 2008)

Theije M De and Heemskerk M, 'Moving Frontiers in the Amazon: Brazilian Small-Scale Gold Miners in Suriname' [2009] European Review of Latin American and Caribbean Studies/Revista Europea de Estudios Latinoamericanos y del Caribe 5

Thielborger P, 'Re-Conceptualizing the Human Right to Water: A Pledge for a Hybrid Approach' (2015) 15 Human Rights Law Review 225 <http://hrlr.oxfordjournals.org/cgi/doi/10.1093/hrlr/ngv008>

Thielbörger P, 'The Right to Water: Effective Multi-Level Protection of a Multi-Faceted Human Right? – An Application of the Kadi and Medellin Approaches to the Case of the Right to Water'

Thielbörger P, 'Philosophical and Conceptual Approaches to a Human Right to Water', *The Right(s) to Water* (Springer Berlin Heidelberg 2014) <http://link.springer.com/10.1007/978-3-642-33908-0> accessed 12 March 2014

Thielbörger P, *The Current Legal Status of the Right to Water* (Springer Berlin Heidelberg 2014) <http://link.springer.com/10.1007/978-3-642-33908-0> accessed 12 March 2014

Thielbörger P, *The Right(s) to Water The Multi-Level Governance of a Unique Human Right* (Springer Science+ Business Media BV 2014)

THIGL, 'The Political Economy of Water Management in Yemen' (2014)

Tladi D, 'Intragenerational Equity : A New Name for International Environmental Justice' (2003) 9 Freedom, Justice and Equality : three Pillars of Legal History 197

Tladi D, *Sustainable Development in International Law : An Analysis of Key Enviro-Economic Instruments* (2007)

Tortajada C, 'IWRM Revisited: From Concept to Implementation' (2014) 30 International Journal of Water Resources Development 361 <www.tandfonline.com/doi/abs/10.1080/07900627.2014.937085> accessed 28 August 2014

Tremblay H, 'A Clash of Paradigms in the Water Sector ? Tensions and Synergies Between Integrated Water Resources Management and the Human Rights-Approach to Development' (2011) 51 Natural Resources Journal 307

Trigueros A, 'The Human Right to Water : Will Its Fulfillment Contribute to Environmental Degradation ? The Human Right to Water : Will Its Fulfillment Contribute to Environmental' (2012) 19 Indiana Journal of Global Legal Studies 599

Trouwborst A, 'Precautionary Rights and Duties of States' (Utrecht University 2006) <http://papers.ssrn.com/sol3/papers.cfm?abstract_id=2556393%0Ahttp://dspace.library.uu.nl/bitstream/handle/1874/10017/full.pdf?sequence=16>

Trouwborst A, 'The Precautionary Principle in General International Law: Combating the Babylonian Confusion' (2007) 16 Review of European Community and International Environmental Law 185

Tully S, 'A Human Right to Access Water? A Critique of General Comment No. 15' (2005) 23 Netherlands Quarterly of Human Rights 35

Turner SJ, *A Global Environmental Right* (Routledge 2014)

Ulrich MR, 'The Impact of Law on the Right to Water and Adding Normative Change to the Global Agenda' 48 George Washington International Law Review 43

UN, 'The Sustainable Development Goals Report 2016' (2016) <www.theguardian.com/global-development/ng-interactive/2015/jan/19/sustainable-development-goals-changing-world-17-steps-interactive>

UN-Water, 'Eliminating Discrimination and Inequalities in Access to Water and Sanitation' (2015)

UN-Water, *The United Nations World Water Development Report 2015: Water for a Sustainable World* (UNESCO 2015)

UN Committee on the Elimination of Discrimination Against Women, 'General Recommendation No. 34 on the Rights of Rural Women' 1

UNEP, 'Vulnerability of People and the Environment – Challenges and Opportunities' (J Jager and Kok eds, 2007) <www.jci.org/cgi/content/abstract/115/10/2656%5Cnwww.tandfonline.com/doi/full/10.3109/09553002.2015.1047989%5Cnwww.pubmedcentral.nih.gov/articlerender.fcgi?artid=3820435&tool=pmcentrez&rendertype=abstract%5Cnwww.nature.com/doifinder/1>

UNEP, *Option for Decoupling Economic Growth from Water Use and Water Pollution* (UNESCO 2015)

UNESCO, *Managing Water under Uncertainty and Risk*, vol 1 (UNESCO 2012)

UNICEF, 'Suriname Multiple Indicator Cluster Survey 2010' (2013)

UNICEF and WHO, 'Progress on Drinking Water and Sanitation 2013 – Update' (2013)

United Nations, 'World Water Development Report (Executive Summary)' (2003)

United Nations, 'Women and Water' (2005)

United Nations, 'Guiding Principles on Business and Human Rights Guiding Principles on Business and Human Rights' 35

United Nations Development Programme, *Human Development Report 2006 Beyond Scarcity: Power, Poverty and the Global Water Crisis* (2006) <www.undp.org/content/dam/undp/library/corporate/HDR/2006 Global HDR/HDR-2006-Beyond scarcity-Power-poverty-and-the-global-water-crisis.pdf>

Van Bardeleben M, *Implementing Sustainability* (2011)

van der Valk M and Keenan P (eds), 'The Right to Water and Water Rights in a Changing World', *Colloquium the right to water and water rights in a changing world* (2011)

Van Rijswick H. FMW, 'Improving the Right to Water in the Netherlands' in Henri Smets (ed), *The implementation of the right to safe drinking water and sanitation in Europe* (2011)

van Rijswick M and others, 'Ten Building Blocks for Sustainable Water Governance: An Integrated Method to Assess the Governance of Water' [2014] Water International 1 <www.tandfonline.com/doi/abs/10.1080/02508060.2014.951828> accessed 8 September 2014

Vink K, 'Transboundary Water Law and Vulnerable People: Legal Interpretations of the "equitable Use" Principle' [2014] Water International 1 <www.tandfonline.com/doi/abs/10.1080/02508060.2014.951827> accessed 16 September 2014

Vink K, Takeuchi K and Kibler KM, 'A Quantitative Estimate of Vulnerable People and Evaluation of Flood Evacuation Policy' (2014) 9 Journal of Disaster Research 887

Voigt C, *Sustainable Development as a Principle of International Law* (Martinus Nijhof Publishers 2009)

Water Governance Facility, 'Human Rights-Based Approaches and Managing Water Resources: Exploring the Potential for Enhancing Development Outcomes' (2012)

WaterAid, 'Rights-Based Approaches to Increasing Access to Water and Sanitation' (2011)

Waterlex; and WASH United, 'The Human Rights to Water and Sanitation in Courts Worldwide' (2014)

Webley L, 'Qualitative Approached to Emperical Legal Research' in Peter Cane and Herbert Kritzer (eds), *The Oxford Handbook of Empirical Legal Research* (Oxford University Press 2010)

Webster TC and Roebuck L, 'Water Resources Assessment of Suriname' (2001)

Wei Z (ed), *Human Rights and Good Governance* (Brill Nijhoff 2016)

Weiss EB, 'The Planertary Trust: Conversation and Intergenerational Equity' (1984) 11 Ecology Law Quarterly 495

Weiss EB, 'Intergenerational Equity : A Legal Framework for Global Environmental Change' in Edith Brown Weiss (ed), *Environmental change and international law: New challenges and dimensions* (United Nations University Press 1992)

Westra L, 'Climate Change and the Human Right to Water' (2010) 1 Journal of Human Rights and the Environment 161 <http://e-elgar.metapress.com/openurl.asp?genre=article&id=doi:10.4337/jhre.2010.02.02>

White G, Bradley D and White A, 'Drawers of Water' (1972) 80 Bulletin of the World Health Organization 63 <www.who.int/entity/bulletin/archives/80(1)63.pdf>

WHO/UNICEF, 'Drinking Water Equity, Safety and Sustainability. Thematic Report on Drinking Water' (2011)

WHO/UNICEF, '25 Years Progress on Sanitation and Drinking Water' (2015)

Wibisana A, 'Three Principles of Environmental Law : The Polluter-Pays Principle, the Principle of Prevention, and the Precautionary Principle', *Environmental Law in Development: Lessons from the Indonesian Experience* (Elgar Publishers 2006)

Williams LA, 'Resource Questions in Social and Economic Rights Enforcement: A Preliminary View', *Social and Economic Rights in Theory and Practice: A Critical Inquiries* (2015)

Winkler I, 'Judicial Enforcement of the Human Right to Water – Case Law from South Africa, Argentina and India' (2008) 1 Law, Social Justice & Global Development Journal 2

Winkler I, 'Policy Guidance Workshop: The Human Right to Water and Sanitation "Translating Theory into Practice"' (2008)

Winkler I, *The Human Right to Water Significance, Legal Status and Implications for Water Allocation* (First, Hart Publishing 2012)

Winkler I, 'The Human Right to Sanitation' (2016) 37 University of Pennsylvania Journal of International Law 1331 <http://papers.ssrn.com/abstract=2728068>

Winkler IT, *The Human Right to Water* (2015)

Wip D and others, 'Urban Mercury Pollution in the City of Paramaribo, Suriname' (2013) 6 Air Quality, Atmosphere and Health 205

Woodhouse M and Langford M, 'There Is No Human Right to Water for Livelihoods' (2009) 28 Waterlines 5 <http://developmentbookshelf.com/openurl.asp?genre=article&id=doi:10.3362/1756-3488.2009.002> accessed 6 March 2013

Woods K, *Human Rights and Environmental Sustainability* (Edward Elgar Publishing 2010)

World Business Council for Sustainable Development, 'Facts and Trends: Water' (2006)

World Health Organization, 'Mercury in Drinking-Water. Background Document for Development of WHO Guidelines for Drinking-Water Quality' (2005)

World Health Organization, *Protecting Groundwater for Health Managing the Quality of Drinking-Water Sources WHO Drinking-Water Quality Series Protecting Groundwater for Health. Managing the Quality of Drinking-Water Sources* (2006)

World Health Organization, 'Guidelines for Drinking-Water Quality' (2011)

World Health Organization, 'GLAAS 2012 Report. UN-Water Global Analysis and Assessment of Sanitatioin and Drinking-Water' (2012) <www.un.org/waterforlifedecade/pdf/glaas_report_2012_eng.pdf>

Ziganshina D, 'Rethinking the Concept of the Human Right to Water' (2008) 6 Santa Clara Journal of International Law 113

Agenda 21 1992 1

General Comment No. 14 The right to the highest attainable standard of health 2000 1

General Comment No. 4: The Right to Adequate Housing (Art. 11) 1999 1

General Comment No. 6: Article 6 (Right to Life) 1982 1

Human rights and access to safe drinking water and sanitation: resolution adopted by the Human Rights Council 2010 4

International Covenant on Economic, Social, and Cultural Rights 2009 1

APPENDIX

LIST OF EXPERTS AND STAKEHOLDERS WHO PARTICIPATED IN INTERVIEWS

Semi-structured interviews took place with the below mentioned experts and stakeholders.

1.	Name	Position
2.	Guido van der Kooye	Coordinator Ministry of Agriculture, Fishery and Animal Husbandry Region West (Nickerie)
3.	Richenel Small	Acting director OWMCP
4.	Christopher Healy	Manager Culturecom Consulting, Antropologist
5.	Alwin Linger	Vice-president Operations N.V. Surinaamse Waterleiding Maatschappij
6.	Mariska Riedewald	Commissioner Environment (Ministery of Labour, Technology and Environment)
7.	Humphrey R. Schurman	Schurman Lawyers
8.	Margo Waterval	University Professor Law Department Anton de Kom University Suriname
9.	Otto Ferf Jentink	Policy-advisor Wereld Waternet
10.	Marieke Heemskerk	Consultancy Firm Social Solutions
11.	Farzia Hausil	Legal advisor NIMOS (Nationaal Instituut voor Milieu en Ontwikkeling in Suriname – National Institute for Environment and Development)
12.	Gina Griffith	Legal advisor NIMOS
13.	Bryan Drakenstein	UNDP Suriname Specialist Energy & Environment
14.	Ruben Ravenberg	Regional Manager East at Surinaamsche Waterleiding Maatschappij N.V.
15.	Mascha Baak	Embassy of the Netherlands Kingdom in Paramaribo policy focus water and climate
16.	Anonymous	Judge at Court in Paramaribo
17.	Rita Henry	Head Legal Department International and Legal Affairs Ministry of Labour, Technological Development and Environment
18.	John Goedschalk	Director Conservation International
19.	Manodj Hindori	Former Chairman Overliggend Waterschap MCP (OWMCP) Head of WaterForum Foundation Suriname (WFS)
20.	Gloria Sandi	Lawyer Surinaamse waterleiding maatschappij

	Name	Position
21.	Riad Nurmohamed	Lecturer/Researcher/Head of Department ADEK
22.	Ms. Cheryl Sastro	Assistant Coordinator (Water and Sanitation Department) N.V.B. Stichting Nationale Vrouwen Beweging (National Women's Movement)
23.	Nancy del Prado	Former Operations Support Consultant at Inter-American Development Bank Environmental Law and Policy Consultant
24.	Chiquita Resomardono	Head of Watercooperation
25.	Sieuwnath Naipal	Hydrologist at ADEK
26.	Ms. Jornell Vinkwolk	Coordinator Human Rights Bureau of Ministry of Justice & Police
27.	Harvey Bijnaar	Environmental inspector
28.	Ms. Meryll Malone	Senior Desk Officer Multilateral Affairs Division Ministry of Foreign Affairs
29.	Thanya Soke-Fonkel	Vice-director Development of the Interior Ministry of Regional Development
30.	Johanna Lakhisaran	WHO/PAHO Technical Officer at Pan American Health Organisation / World Helath Organisation
31.	Haidy Malone	UNDP
32.	Henry Vreedzaam	Small grants
33.	Revinh Ramnandanlal	Policy Advisor, Change Management & Environmental Issues at Cabinet of the Vice President of the Republic of Suriname
34.	Ravi Patandin	Engineering firm Ilaco Suriname, WaterForum
35.	Wilco Finisie	Director Ministry of Regional Development
36.	Negretha Amiemba	International Relations assistant Ministry of Regional Development
37.	Donnevan Eduards	SWM manager in Republiek

Human Rights Research Series

The Human Rights Research Series' central research theme is the nature and meaning of international standards in the field of human rights, their application and promotion in the national legal order, their interplay with national standards, and the international supervision of such application. Anyone directly involved in the definition, study, implementation, monitoring, or enforcement of human rights will find this series an indispensable reference tool.

The Series is published together with the world famous Netherlands Network for Human Rights Research (formerly School of Human Rights Research), a joint effort by human rights researchers in the Netherlands.

For previous volumes in the series, please visit http://shr.intersentia.com.

Published titles within the Series:

73. Roland Moerland, *The Killing of Death* ISBN 978-1-78068-351-5
74. Andrea Broderick, *The Long and Winding Road to Disability Equality and Inclusion* ISBN 978-1-78068-358-4
75. Christina Peristeridou, *The Principle of Legality in European Criminal Law* ISBN 978-1-78068-357-7
76. Emilie Kuijt, Humanitarian Assistance and State Sovereignty in International Law ISBN 978-1-78068-366-9
77. Lize R. Glas, *The Theory, Potential and Practice of Procedural Dialogue in the European Convention on Human Rights System* ISBN 978-1-78068-375-1
78. Paulien de Morree, *Rights and Wrongs under the ECHR* ISBN 978-1-78068-418-5
79. Malu Beijer, *The Limits of Fundamental Rights Protection by the EU* ISBN 978-1-78068-455-0
80. Marie Elske Gispen, *Human Rights and Drug Control* ISBN 978-1-78068-454-3
81. Bart van der Sloot, *Privacy as Virtue: Moving Beyond the Individual in the Age of Big Data* ISBN 978-1-78068-505-2
82. Mohammad Hadi Zakerhossein, *Situation Selection Regime at the International Criminal Court* ISBN 978-1-78068-618-9
83. Yi Zhang, *Advancing the Right to Health Care in China. Towards Accountability* ISBN 978-1-78068-705-6
84. Diana Itza Odier-Contreras Garduño, *Collective Reparations.Tensions and Dilemmas Between Collective Reparations and the Individual Right to Receive Reparations* ISBN 978-1-78068-705-6

Lightning Source UK Ltd.
Milton Keynes UK
UKHW052002220219
337857UK00004B/128/P

9 781780 686127